ENCOUNTERS WITH THE UNKNOWN

A JOURNEY THROUGH SECRETS,
SCIENCE, AND THE HUMAN SKY

DR. CARL RYAN TUCKER

ISBN: 979-8-218-94359-2

ABOUT THE AUTHOR

Dr. Carl Ryan Tucker is an author, scholar, and IT executive who moves comfortably between complex systems and the human stories inside them. His work blends rigorous research, practical leadership experience, and a healthy skepticism that questions easy answers while searching for what is true, useful, and humane. Across topics that range from information technology and leadership to metaphysics, spirituality, and the unknown, Dr. Tucker invites readers to think critically, stay curious, and apply what they learn to real life. His writing is grounded in evidence, shaped by practice, and fueled by the belief that knowledge should empower people to build more resilient, ethical, and imaginative futures.

PREFACE

This book didn't begin with a dramatic sighting or a single moment of revelation. It began with a quiet realization that refused to go away. Ufology has always lived between science, culture, and politics, inviting curiosity and provoking skepticism. For decades, Unidentified Aerial Phenomena have been dismissed as fringe or absorbed into entertainment. Yet, beneath that noise lies something more consequential. UAP sit at the intersection of technology, secrecy, and imagination, exposing the limits of what we think we understand.

What started as an effort to trace the history of UAP research quickly expanded. The Cold War revealed how secrecy, classification, and misdirection shaped public understanding, from Project Sign and Blue Book to AATIP, UAPTF, and AARO. The pattern was consistent. Denial followed by partial disclosure, curiosity constrained by defense priorities. Looking beyond the United States added clarity. France's GEIPAN, Brazil's Operation Saucer, and Chile's CEFAA showed how political culture determines whether the unknown is treated as a scientific problem or something to manage and contain. Across these cases, UAP reflected not only anomalous events but the fears and priorities of the societies studying them.

As the research deepened, the human element became impossible to ignore. Witness accounts were shaped by perception, belief, and media narratives. Psychologists pointed to cognitive bias. Sociologists examined how stigma polices the boundaries of acceptable inquiry. UAP emerged as both external anomalies and internal stories, shaped as much by the human mind as by physics or atmospheric conditions.

At the same time, the private and civilian sectors began to fill gaps left by government hesitation. Organizations like MUFON, SCU, and UAPx combined volunteer reporting with technical analysis. Aerospace firms, data platforms, and citizen science projects introduced new tools, from machine learning to open sensor networks. Ufology began to resemble a collaborative field rather than a fringe pursuit.

This book is not about proving extraordinary claims. It is about understanding how institutions, cultures, and individuals respond when confronted with the unexplained. It follows pilots, engineers, researchers, and witnesses as they struggle to describe events that resist easy categories. At its core, this is a study of integration. Progress depends on perspectives working together rather than competing for authority.

The story of UAP is ultimately a story about uncertainty and how humans live with it. Every era reshapes the phenomenon in its own image. My hope is that readers leave with curiosity sharpened by discernment, and skepticism balanced by openness. The mystery matters less than how we choose to engage with it.

TABLE OF CONTENTS

PART I

THE AGE OF SECRETS

"The UFO phenomenon calls for a scientific curiosity unfettered by dogma."

- J. Allen Hynek -

When James La Pierre began tracing the history of Unidentified Aerial Phenomena, he realized the mystery was never confined to the sky. It lived inside institutions. In how militaries observe anomalies, how governments decide what may be known, and how uncertainty is managed when explanations fail. Part I follows the birth of modern ufology in the late 1940s, when Cold War anxiety, technological surprise, and public fascination collided. From that moment on, UAP became less a question about what was seen and more a test of how power responds to the unexplained.

The story opens in 1947 with Kenneth Arnold and Roswell, then moves through decades of secrecy shaped by intelligence rivalry and defense priorities. Programs like Sign, Grudge, and Blue Book reveal a pattern where strategic control often eclipsed scientific curiosity. That pattern never vanished but adapted, reemerging through AATIP, the UAP Task Force, and AARO amid whistleblowers, congressional pressure, and international contrast. Part I closes by confronting the deeper problem beneath the sightings. UAP expose not only limits in sensor technology, but limits in how institutions define knowledge itself.

THE SUMMER THE SKY CHANGED

"Something in the sky was seen, and it made people think differently about the world."

- Carl Jung -

A Man and the Question Above

The hum of the projector filled the dim library room where James La Pierre sat alone, watching the grainy black-and-white footage flicker across the screen. Tiny lights drifted through the night sky, distant, uncertain, and unresolved. The film had been shot in 1952, yet for James, it felt immediate. The objects seemed alive, unbothered by human interpretation.

He was not a conspiracy theorist, nor a man given to easy belief. Trained as a historian with a background in science, James approached the subject of Unidentified Aerial Phenomena with the same careful skepticism he brought to any archival mystery. But something about the footage tugged at him. The way people in the film spoke, nervous, reverent, and hopeful, suggested something more than a simple misunderstanding of weather or radar. It suggested awe.

And so began his journey into ufology. The study of the unknown in the skies, a field suspended between science and story with fact and faith.

The World Before Flying Saucers

Before 1947, tales of strange things in the heavens were not uncommon. James spent weeks combing through microfilm in old university archives, finding accounts from the 1890s of “mystery airships” gliding over the American Midwest. These reports, printed in regional newspapers, spoke of cigar-shaped craft and searchlights that seemed to dance among the stars. Most historians dismissed them as hoaxes or misidentifications inspired by the era’s fascination with flight. But James noticed something deeper. The pattern of imagination that would echo across the next century.

People had always projected their anxieties and aspirations into the sky. The invention of flight had simply given those projections shape. Each generation saw in the heaven’s whatever technology, fear, or hope defined its age. In the nineteenth century, the airship symbolized innovation. By the twentieth, it would become a mirror for global paranoia.

1947: The Summer the Sky Changed

James first encountered Kenneth Arnold’s original interview transcript in a faded folder at the National Archives. The man’s words leapt from the page with disarming sincerity. On June 24, 1947, Arnold, a civilian pilot, had been flying near Mount Rainier when he saw nine gleaming objects skipping across the sky “like saucers on water.” Within days, the Associated Press had seized upon the term “flying saucers,” embedding it permanently in popular vocabulary.

Arnold’s report was soon followed by hundreds of others. Across the country, from farmers to radar operators, witnesses described objects that defied explanation. Newspapers ran sensational headlines. America was transfixed. For James, it was clear that this was not merely an outbreak of mistaken perception. It was a cultural event. A sudden collision of technology, imagination, and fear in the atomic age.

The frenzy reached its crescendo two weeks later in Roswell, New Mexico. James traced the sequence of telegrams exchanged that July between military bases. The first announced the recovery of a "flying disc." The next day's telegram retracted the claim, citing a weather balloon. The speed of reversal was stunning, and the damage irreversible. What had begun as a straightforward press release became, through contradiction, a myth of secrecy that would outlive everyone involved.

When James visited Roswell himself decades later, he found a town that had turned mystery into identity. Gift shops sold "alien jerky" and green-skinned dolls. Yet beneath the kitsch lay something haunting. A collective sense that truth had once been glimpsed, then buried. He realized that Roswell had become less a story about aliens and more a parable about trust.

Cold War Shadows and the Machinery of Secrecy

The deeper James went into military archives, the more he saw how the Cold War shaped the birth of ufology. In 1948, the U.S. Air Force launched Project Sign, followed by Grudge and Blue Book, programs tasked with determining whether these sightings represented a threat. Declassified memos revealed an early openness to the extraterrestrial hypothesis, quickly replaced by bureaucratic skepticism after the CIA's 1953 Robertson Panel advised debunking UFOs to prevent public hysteria.

James imagined the tension in those rooms, scientists and officers poring over radar traces, wondering if the Soviets had developed a secret craft or if the phenomena came from even farther afield. But official statements soon hardened. UFOs, once treated as a possible breakthrough, became framed as a nuisance.

Still, the sightings continued. High-altitude reconnaissance programs like the U-2 and A-12 Oxcart, both classified, produced countless false UFO reports. Pilots saw strange lights that even base commanders couldn't identify because the projects were compartmentalized. Ironically, American secrecy itself fed the very rumors it sought to quell.

For James, this feedback loop between hidden technology and public imagination was the true engine of the UFO era. Each classified flight spawned whispers and each denial confirmed suspicion. It was less about what people saw than about what they were not allowed to know.

Science, Skepticism, and the People Who Looked Up

When James attended a small UFO conference in Arizona, he expected wild claims. Instead, he found a curious mix of amateur scientists, retired pilots, and ordinary citizens driven by the same question that had drawn him in. One man, a former radar operator from the 1950s, told him, "We weren't crazy. We just didn't have clearance."

Mainstream scientists, however, largely avoided the subject. Academic journals dismissed UFO reports as anecdotal. The stigma was strong enough that researchers who showed interest risked professional exile. James saw this divide as central to the story. A society obsessed with rationality confronting experiences that refused easy categorization.

Psychological studies later confirmed what James observed firsthand. Human perception is fallible. Our brains impose patterns, fill gaps, and favor meaning over randomness. Yet the persistence of belief in UFOs was not irrational. It was profoundly human. It reflected the desire to understand mystery within a world increasingly dominated by machines and data.

The Roswell Feedback Loop

In his notebooks, James sketched the Roswell story like an ouroboros, the snake devouring its own tail. Each time the government explained the incident, new inconsistencies appeared, feeding further doubt. The 1990s declassification of Project Mogul balloon tests was supposed to end speculation, but to many, it only proved that the truth had been hidden for decades.

James came to see Roswell not as a question of what happened, but as an allegory for how belief operates. In the vacuum of transparent information, imagination rushes to fill the void. Conspiracy theories, psychologists argue, thrive where trust erodes. Once secrecy exists, every disclosure looks suspect, every revelation incomplete.

He thought of Jung's words about UFOs as "modern myths of things seen in the sky," symbols of humanity's collective anxiety. Roswell, in that sense, was the primal myth of modern America. It was a story about how power conceals and how wonder persists.

A Mirror to Modern Society

By the time James reached the 1960s in his research, ufology had matured into a movement. Civilian organizations like the Aerial Phenomena Research Organization and NICAP had built networks of volunteers collecting witness accounts, photographs, and radar data. It was a community living on the margins of science but organized with scientific ambition.

James spent an afternoon with one of the last surviving members of NICAP's early days. The man, frail but sharp, described stuffing newsletters into envelopes by hand, convinced they were part of a great awakening. "We weren't chasing aliens," he said. "We were chasing the truth the government refused to look at."

The Cold War had blurred the line between secrecy and paranoia. When the Vietnam War and Watergate later exposed real deceptions, ufology's mistrust of authority no longer seemed fringe. It seemed prescient. James saw how UFO mythology quietly shaped American skepticism, merging with broader movements demanding accountability and transparency.

In pop culture, meanwhile, the phenomenon evolved from menace to metaphor. Films of the 1950s portrayed UFOs as invasion threats. By the 1970s, they had become symbols of transcendence and hope. The alien shifted from conqueror to savior, mirroring society's changing relationship with technology and belief.

Standing Beneath the Stars

Late one night, James drove into the Nevada desert, far from light pollution, and parked his car beneath the vast dome of stars. The silence was immense. He thought about all the people who had stood on lonely roads like this one, scanning the sky for answers, meaning, or proof. Anything!

Whether the lights they saw were aircraft, illusions, or something beyond, what mattered most, he realized, was the human response. The stories, the yearning, the persistence of wonder in an age that so often dismissed it. Ufology was not merely about unidentified objects. It was about curiosity, fear, awe, and distrust, all orbiting around the same enduring mystery.

The Sky as a Mirror

Back in his study, surrounded by boxes of declassified documents and yellowed newsprint, James understood why ufology had endured despite ridicule. It was not simply a fringe pursuit. It was a mirror held up to modern civilization. In its blend of skepticism and belief, it revealed how humans negotiate uncertainty in a world defined by secrecy and science.

From Arnold's shimmering discs to Roswell's shifting stories, from Cold War intelligence to the modern Navy's "Tic Tac" videos, the phenomenon had evolved, yet the question remained timeless.

James turned off the projector, and the room fell into darkness. Outside the window, the night sky stretched wide and silent. Somewhere out there, or perhaps only within the human mind, the unknown still waited, gleaming just beyond explanation.

THE POLITICS OF THE UNKNOWN

"Secrecy is the beginning of tyranny."

- Robert Heinlein -

The Hearing Room

The hearing chamber buzzed like a hive. Cameras hummed softly. Reporters whispered names as though reciting a litany. *Fravor, Graves, Grusch.* The date was July 26, 2023, and for the first time in modern memory, Unidentified Aerial Phenomena (UAP) had taken center stage in the United States Capitol.

James La Pierre sat among the journalists and policy analysts in the public gallery, notebook open, pen balanced across his knee. He had spent years tracing the government's tangled relationship with the unknown in the skies, but seeing the faces of Navy pilots sworn in under oath made the mystery feel suddenly solid.

Commander David Fravor, silver-haired and calm, described an object shaped like a "Tic Tac," dropping from thirty thousand feet to sea level in seconds. Beside him, Ryan Graves described encounters off the East Coast. A craft without exhaust plumes or visible flight surfaces, moving through restricted airspace as if the laws of physics had momentarily stepped aside.

When Representative Ogles asked whether the pilots had ever been briefed about recovered materials or "non-human biologics," the room's oxygen seemed to vanish. That was the line between rumor and revelation. And then David Grusch, the whistleblower who had electrified Washington, leaned toward the microphone and said, *"Yes, I have been told of such programs."*

James exhaled, his pen frozen above the page. For months he had sifted through declassified memos, FOIA releases, and whispered interviews in cafés near the Pentagon. He knew the pattern by heart. Secrecy, denial, partial disclosure, more secrecy. Yet hearing it declared aloud in Congress gave the decades-long paper trail a human heartbeat.

He wondered, as cameras flashed and the chairman called for order, how a nation built on transparency had turned the heavens into a classified file.

Shadows of the Cold War

In the months following that hearing, James retraced the story to its beginning. The archive room at Wright-Patterson Air Force Base still smelled faintly of dust and jet fuel. Here, under buzzing fluorescent lights, he read the old memos stamped **PROJECT SIGN - TOP SECRET**.

In 1948, military intelligence officers had gathered to analyze a flood of UFO reports sparked by Kenneth Arnold's sighting the previous summer. The newly formed U.S. Air Force was desperate to know whether Soviet engineers had leapt ahead in aeronautics. Sign's analysts catalogued shapes, speeds, and altitudes, occasionally venturing to speculate about "interplanetary origins."

Then came Project Grudge, 1949. The tone turned colder. Memos warned against "unscientific enthusiasm." Reports were filed away under "psychological misinterpretation." The bureaucratic pivot was clear. Mystery threatened order and order was the currency of the Cold War.

James pictured the analysts, men in uniform shirtsleeves with coffee rings staining classified reports, arguing late into the night over whether a glowing disc above Idaho was the planet Venus or a Soviet spy craft. To them, the question wasn't cosmic but was tactical. Every unidentified radar blip could be a Soviet test flight, a balloon, or an opportunity for panic.

By 1952, Project Blue Book formalized the effort. Over the next seventeen years it logged more than twelve thousand cases. James studied the case files. There were sketches drawn by farmers, radar returns annotated by colonels, and letters from civilians demanding the truth. Captain Edward Ruppelt, Blue Book's first director, had tried to keep an open mind, but the system valued plausible deniability more than discovery.

The Cold War mindset hung over every paragraph. *Control the narrative and contain the fear.*

The Curtain of Silence

When Blue Book shut down in 1969, headlines announced that the U.S. Air Force had officially lost interest in UFOs. But the documents James found told another story.

In a windowless reading room in College Park, Maryland, he examined a sheaf of declassified CIA memos. One noted that "unidentified radar contacts over the Arctic" should be cross-referenced with Soviet satellite launches. Another mentioned "psychological operations potential." The signature at the bottom, smudged and unreadable yet unmistakably real, was proof that curiosity had not vanished at all. It had merely gone underground.

There was no single agency now. The only fragments that remained were the NSA scanning frequencies, NORAD watching blips, and the CIA correlating reports with spy-plane test routes. The public silence was strategic. Better to let the UFO craze burn itself out than to admit that the military didn't always know what filled its skies.

As James drove away from the archive that day, he thought of the strange irony. A democracy hiding its ignorance behind secrecy. The citizens wanted certainty and the state wanted calm. Between them stretched a fog thick enough to swallow truth whole.

The Quiet Resurrection

Nearly four decades later, in 2007, a classified budget line revived official curiosity. James learned about it through a Senate staffer who spoke in riddles over black coffee in Arlington. The name was innocuous. *Advanced Aerospace Threat Identification Program* (AATIP).

It was born inside the Defense Intelligence Agency with a modest budget and a mandate to study unconventional flight phenomena. The funding came through Senator Harry Reid, whose handwritten note in a released letter called for "anomalous aerospace threats research."

James traced the contracts to a Nevada aerospace firm exploring metamaterials and "space-time metric engineering." To some, it sounded visionary, but to others, fringe. But the program's real spark came from Navy pilots reporting encounters they couldn't explain.

For nearly a decade, AATIP operated quietly in the shadows, its findings sealed away and its personnel known only to a select few. Then in 2017, *The New York Times* broke the story. Alongside it came three Navy videos. They were blurry, infrared, yet undeniable. Objects with no visible propulsion darting across the screen while astonished pilots cursed in disbelief.

James remembered watching those clips the night they first aired. What startled him wasn't the footage itself, but the acknowledgment that followed. The Pentagon confirmed the videos were authentic. For the first time since Project Blue Book, the government had acknowledged something it couldn't explain.

The Return to Daylight

By 2020, the Pentagon could no longer dismiss the growing pile of encounters. The Unidentified Aerial Phenomena Task Force was established within Naval Intelligence. Its mission was to standardize reports, brief Congress, and assess potential threats.

James interviewed a retired Navy analyst who described it as "catch-up after seventy years." He spoke carefully, weighing each word. "The data's a mess," he said. "Different services, different formats. We needed one language for the unknown."

In 2021, the Office of the Director of National Intelligence released its Preliminary Assessment. It catalogued 144 cases, with only one explained. The report was cautious and almost clinical, but its subtext was electric. The government no longer ridiculed UAP. It studied them.

A year later, under congressional pressure, the Department of Defense created the All-domain Anomaly Resolution Office (AARO). Its logo, newly printed on briefing slides, showed Earth encircled by rings of air, sea, and space. For James, that image felt like a confession. The mystery had grown beyond the skies.

He attended AARO's first press briefing in 2024. The director spoke in polished bureaucratese about "trans-medium objects" and "data integrity." What struck James most was the tone itself. It was careful, deliberate, and undeniably genuine. They were speaking publicly about something the government once refused to name.

That same year, AARO launched a public reporting portal. Civilians could now submit sightings directly to an official database. It wasn't full disclosure, but it was daylight after decades of dusk.

The Whistleblower and the Firestorm

By the summer of 2023, the story that had simmered for seventy years boiled over. David Grusch, a decorated intelligence officer, filed a whistleblower complaint alleging that the U.S. government possessed retrieved craft "of non-human origin."

James met him briefly at a closed-door forum months before the congressional testimony. Grusch spoke quietly, with the measured cadence of someone used to briefings, not headlines. "I'm not asking people to believe," he said. "I'm asking them to investigate."

When his statements became public, Washington split down predictable lines. Skeptics demanded evidence. Others whispered that Grusch had said what insiders had hinted for decades. The Pentagon issued a careful denial. *No verifiable information supports the claims.*

James watched the reactions spread outward, journalists dissecting every word while lawmakers hurried to draft new transparency bills. The most prominent was the UAP Disclosure Act of 2023, modeled after the JFK Records Act, seeking to mandate systematic declassification of UAP files.

Even if the bill failed to pass intact, it symbolized a shift that the era of quiet denial was ending.

For James, Grusch's testimony wasn't proof of aliens. It was evidence of a fracture, a fault line where secrecy met democracy.

Across the Atlantic and Beyond

To widen his lens, James flew to London. At the British National Archives in Kew, he examined the once-classified UFO files released between 2008 and 2013. Thousands of pages detailed civilian and police sightings, each stamped "DEF 23/497." The tone of the memos was weary but transparent. The Ministry of Defence had closed its UFO desk in 2009, declaring the matter of "no defense significance." Yet even here, patterns began to appear, with blips tracked on radar and unexplained lights seen over airbases.

Next, he traveled to Toulouse, where France's GEIPAN program operated inside the national space agency CNES. The office felt more like a university lab than a military post. Scientists discussed classification frameworks, atmospheric data, and witness psychology. "Transparency is our insulation against hysteria," one researcher told him. "When the public can read our files, they trust the process."

In Brazil, James met retired officers who had served during Operation Saucer in the late 1970s. They spoke of luminous orbs above the Amazon and villagers injured by beams of light. Decades later, the Brazilian Air Force released those documents, a rare act of transparency within Latin America. One colonel admitted, "We didn't know what it was, but we knew hiding it would only make it worse."

Each country's approach reflected its culture. Britain closed the file. France kept it open. Brazil shared the story. The

United States, James realized, had done all three, though never within the same decade.

The Machinery of Secrecy

Back home in Virginia, James began assembling his notes into a timeline stretching from 1947 to 2024. Redacted pages lay beside glossy congressional reports, their contrast almost poetic. The age of carbon-copied memos giving way to televised hearings.

Yet one pattern persisted. Every attempt at disclosure was also an attempt at control. Blue Book had aimed to reassure. AATIP had aimed to contain. AARO aimed to standardize. James thought, the truth wasn't a hidden object but a moving target, shaped by whoever happened to hold the pen.

He thought often about Heinlein's warning that secrecy breeds tyranny. In intelligence work, secrecy was routine, even necessary. But when secrecy became reflex, it transformed curiosity into suspicion. The longer a mystery stayed classified, the more it grew teeth in the public imagination.

The Open Sky

Late one evening, James returned to the Capitol steps, where the summer air smelled faintly of rain. Looking up, he saw a commercial jet tracing its steady line across the dark. Beyond it, stars pulsed in silence.

The hearings, the archives, and the memos were all human attempts to bring order to that silence. Perhaps, he thought, the real story of government involvement in UAP wasn't about aliens or conspiracies at all. It was about how societies manage uncertainty.

In every decade since 1947, the same paradox had played out. To protect the nation, agencies concealed what they couldn't explain, yet by concealing it, they undermined the trust they sought to preserve.

James closed his notebook. The night above Washington D.C. was clear, infinite, and indifferent. Somewhere in that darkness, whether locked in a classified vault or drifting in the far reaches of space, the unknown persisted.

What mattered now was whether humanity would keep it secret, or finally, simply, look.

GHOSTS ON THE RADAR

"Airspace is sovereignty, and sovereignty must be defended."

- Pentagon Source -

The Briefing

The Pentagon's press auditorium was windowless, the fluorescent light sterile as an operating room. Rows of reporters hunched over tablets while the seal of the Department of Defense glowed blue on the wall behind the lectern.

James La Pierre sat halfway back, notebook balanced on his knee, the hum of air-conditioning steady in his ears. The date was late 2024, the day the All-domain Anomaly Resolution Office (AARO) released its Historical Record Report.

The colonel at the podium spoke in clipped syllables. "We are committed to transparency, to analyzing unexplained aerial phenomena wherever they occur."

The word *phenomena* echoed. Cameras clicked. James jotted a quick note, capturing the tone as cautious and the eyes as restless.

He had spent a decade tracing the hidden history behind those words, through radar logs, declassified memos, and testimonies that oscillated between awe and anxiety. For him, this moment, the official acknowledgment that something unknown had repeatedly entered American airspace, felt less like a revelation and more like a recognition. He knew the story stretched back to radar rooms of the Cold War, to pilots who swore the sky sometimes moved in ways physics had not yet licensed.

The colonel gestured toward a slide showing faint tracks spiraling across radar maps. “These are unresolved detections. No known origin. No hostile intent observed.”

James felt the familiar tension that haunted every government statement on UAP. The need to reassure without admitting ignorance. He knew the silence between the lines was the real subject.

Ghosts on the Radar

Weeks later, James stood in the radar control room of a retired Air Force base in Ohio. The equipment, rescued from decommission, still flickered with a phosphorescent glow. A retired technician named Walter told him stories from 1952, when the radar screens had come alive with impossible blips over Washington, D.C.

“They moved too fast to be aircraft, too straight to be storms,” Walter said, tapping the glass. “We tracked them. Fighters scrambled. Then, gone.”

That summer, the Pentagon created *Project Blue Book*. James had read every surviving file composed of 12,618 cases. Most were explained as weather balloons or misidentified jets, yet a stubborn remainder refused explanation. Inside those few lines of “unresolved,” he saw the outlines of a deeper unease.

The Cold War was young then, the Soviets testing atomic weapons, and radar was the nervous system of the free world. Any unidentified object risked being mistaken for an attack. “Airspace is sovereignty,” one Pentagon memo read, “and sovereignty must be defended.”

James imagined the tension in those rooms, officers debating whether to treat the unknown as a threat or a mirage. Secrecy cloaked both the uncertainty and the pride. Admitting confusion risked panic and claiming certainty risked exposure. The compromise was bureaucracy. Files were stamped *explained* until they weren’t.

The Quiet Years

After Blue Book closed in 1969, the public was told the Air Force had moved on. Yet as James discovered in a declassified CIA memorandum, interest merely went underground. “Unidentified radar returns should be evaluated for foreign origin,” one note advised.

In the archives, James saw the Cold War’s logic unfold. Every radar anomaly was potentially a Soviet satellite, a reconnaissance balloon, or an atmospheric ghost created by the spy-plane programs that even radar operators weren’t cleared to know existed.

He understood now why secrecy bred speculation. The government concealed what it could explain, like classified aircraft, alongside what it could not. To the public, both silences sounded the same.

By the 1980s, civilian groups like MUFON filled the vacuum. Volunteers combed skies the military no longer claimed to watch. The Air Force kept its quiet watch anyway, logging anything that strayed too close to nuclear facilities. “Disinterest,” James wrote in his notes, “was policy, not reality.”

Resurgence in the Shadows

The story jumped forward three decades, to 2007. A single line in a congressional appropriations bill, easy to overlook, quietly allocated $22 million for "aerospace threat identification."

That line birthed the *Advanced Aerospace Threat Identification Program* (AATIP). Hidden inside the Defense Intelligence Agency, it examined reports from Navy pilots describing white, cigar-shaped craft that accelerated without visible thrust.

James interviewed one of AATIP's early analysts in a café outside Arlington. The man stirred his coffee slowly before speaking. "We weren't chasing aliens," he said. "We were chasing data that didn't fit the model."

For ten years, AATIP's work remained classified. Then came December 2017. *The New York Times* published leaked videos from Navy jets. Infrared footage showed small objects darting like living things, pilots exclaiming, *"What is that thing?"*

Public shock reignited congressional curiosity. To James, it felt like the Cold War had returned in digital form with the same astonishment. This time streaming in 4K.

The Modern Network

By 2020, the Department of Defense could no longer manage the growing number of reports through rumor and PowerPoint slides. It created the *Unidentified Aerial Phenomena Task Force.*

James obtained a copy of the ODNI's 2021 Preliminary Assessment. Of 144 cases examined, 143 remained "unexplained." The language was bureaucratic but revolutionary and it admitted ignorance without ridicule.

Two years later, congressional mandates created AARO. Its mission extended across air, sea, and space, covering what officials called "all domains." AARO's analysts met in secure rooms lit by banks of monitors showing radar sweeps, thermal imagery, and satellite telemetry. The age of the lone witness was over. The age of data had begun.

James was allowed into one demonstration through a media liaison. A technician played synchronized feeds of radar plots overlaying infrared traces and electro-optical imagery. For a moment, three sensors aligned on the impossible. An object hovering motionless against sixty-knot winds before disappearing. The technician only shrugged. "Multi-sensor corroboration. Rare, but real."

The Physics Problem

James spent months interviewing radar engineers and fighter pilots. Each described the same riddle. The data looked solid until physics stepped in.

Radar could track speed but not always shape. Thermal sensors found no heat where propulsion should roar. Electro-optical cameras saw light without a source.

In one Navy squadron, a pilot told him, "They move like nothing with mass should move. If they're Russian, we've lost the race. If they're not..."

He let the thought trail off.

The technical language fascinated James, from cross-section returns to Doppler anomalies, but what truly held his attention was the frustration behind it. The military had built a universe of certainty. UAP mocked that certainty.

He learned that modern radar algorithms often discard erratic movements as noise. A phenomenon that defies pattern may literally vanish from the screen because the system is trained to ignore the impossible.

"That's the irony," an engineer told him. "We may be filtering out the unknown by design."

Airspace and Sovereignty

In the hangar of a naval base, James watched a pilot run through flight-sim data showing an encounter off the Virginia coast. The object appeared, dropped 20,000 feet in seconds, and shot away. The pilot turned to James. "If that's a drone, it's one hell of a drone."

The military referred to such cases as strategic unknowns, objects that were neither friendly nor hostile but unclassified and deeply unsettling. Every appearance in restricted airspace undermined deterrence theory. How could you defend sovereignty against something you couldn't identify?

For the Pentagon, the risk wasn't invasion but ambiguity. An adversary could exploit that uncertainty, masking surveillance under the cloak of the unexplained. For commercial aviation, the risk was physical, involving near misses at high altitude and pilots who hesitated to report them out of fear of ridicule.

James obtained FAA correspondence revealing that civilian pilots routinely described "metallic spheres" or "elongated craft" at cruising altitude, yet few filed formal reports. "Career suicide," one captain told him. "No one wants to be the UFO guy."

The gap between what was seen and what was said had become a safety hazard.

The Policy Paradox

In Washington, James followed the paper trail from secrecy to legislation. Congressional committees now demanded briefings. AARO's charter required annual public reports. Yet most data remained classified under national security exemptions.

At a Senate hearing, an aide whispered to him, "It's a tightrope. Reveal too much, and you expose sensor capabilities. Reveal too little, and you feed the conspiracies."

James recognized the pattern of secrecy to protect systems and transparency to protect legitimacy. Neither ever complete.

He learned that AARO's analysts were developing AI-based *sensor fusion* platforms, merging radar, infrared, and optical feeds in real time. Artificial intelligence would, in theory, learn to distinguish anomalies from noise. But even the algorithms struggled when faced with behavior outside known parameters. "Garbage in, garbage out," one scientist quipped. "Except sometimes the garbage moves at Mach 2."

The Global Lens

James expanded his scope beyond the United States.

In London, archivists at the National Archives showed him the Ministry of Defense's "UFO desk" files, tens of thousands of pages that had been released after 2008. The British had treated the phenomenon as public-relations triage. Calm the citizens and close the case. The final memo from 2009 read, "No defense significance."

In Toulouse, at France's space agency CNES, he met scientists at GEIPAN. They spoke of public databases, witness psychology, and plasma physics. "We treat it as an atmospheric problem until proven otherwise," one said, "but we share everything. Transparency is part of defense."

Brazil offered a different story. In Belém, retired officers described Operação Prato, or Operation Saucer, when soldiers in 1977 pursued luminous orbs through the Amazon and recorded injuries among local villagers. Decades later, the Air Force declassified the files. "We feared ridicule more than the lights," one colonel admitted.

To James, the contrast was revealing. Britain buried. France studied. Brazil confessed. Each choice balanced sovereignty, science, and fear.

Lessons of the Unknown

Back in Virginia, James spread his notes across his desk like a cartographer charting uncertainty.

Across eight decades, the pattern was unmistakable. Each generation built better sensors, yet the unknown kept pace. Every improvement in detection yielded new questions rather than answers.

He jotted a conclusion in his notes: *UAP persist because they live at the limits of our instruments and our belief.*

In defense briefings, generals spoke of sovereignty. In laboratories, physicists spoke of data integrity. In cafés, pilots spoke of awe. Each was right in his own vocabulary.

James realized that the government's challenge was not only to identify the objects but also to reconcile the differing languages of the military, science, and politics into a single grammar of the unknown.

The Open Sky

One autumn night, James drove to the Blue Ridge Mountains, far from city glare. He parked on a ridge overlooking the Shenandoah Valley. The Milky Way spilled overhead, clear as code on a dark screen.

He thought of the radar operators of 1952, of AATIP analysts watching blurred infrared shapes, of today's engineers teaching algorithms to recognize what they themselves could not explain.

A plane crossed the horizon, steady and predictable. A clear example of known physics. Yet beyond it, faint points flickered unpredictably and their motion was uncertain.

James smiled faintly. Whether artifact or revelation, the mystery had done its work. It had forced humanity to look harder at its own sky.

For all the discussion of defense, James thought that the real sovereignty at stake might be intellectual, the ability to acknowledge that not everything had yet been understood.

He closed his notebook. The darkness above remained silent, enormous, and wonderfully unresolved.

PART II

THE LABORATORY OF THE SKY

"Extraordinary claims require extraordinary evidence."

- Carl Sagan -

Part II moves from the politics of disclosure into the laboratories, observatories, and control rooms where Unidentified Aerial Phenomena became a scientific and engineering problem. Investigators like James La Pierre confront the limits of measurement, showing how radar, optics, and human perception can transform uncertainty into apparent impossibility. The mystery shifts from what is seen to how seeing works.

This part begins with the physics of observation, examining radar ducting, sensor artifacts, and electro optical distortion, alongside multi sensor fusion systems designed to reduce error. It then turns to contested claims of exotic materials and isotopic anomalies, revealing how extraordinary conclusions often dissolve under controlled analysis. From there, the narrative expands to astrobiology, exoplanet science, and the Fermi Paradox, where disciplined skepticism guards against false discovery. Part II ultimately presents science under pressure. A method tested by secrecy, rarity, and ambiguity, and a reminder that rigor, not wonder alone, determines whether mystery becomes knowledge or mistake.

ANOMALIES AND EQUATIONS

"All great discoveries begin as anomalies."

- Thomas Kuhn -

Into the Lab

The hum of machines filled the Air Force research lab outside Dayton, Ohio. Oscilloscopes blinked amber in the half-light while a technician adjusted a spectrum analyzer tuned to a frequency that, for most people, didn't exist.

James La Pierre leaned against a counter, watching a radar engineer calibrate the antenna array. On the monitor, a trace line flickered between noise, signal, and noise again. "That," the engineer said, "is how the impossible first shows up. As noise you can't quite explain."

It was a fitting metaphor for everything James had set out to understand. UAP, the modern term for what were once called UFOs, had always existed between signal and noise, balanced uneasily between the measurable and the imagined. To understand them, James realized, physics, instrumentation, and humility were required.

The Problem Space

In his travels through military archives and university labs, James kept running into the same paradox. The most dramatic UAP reports described feats that broke the known laws of aerodynamics, yet every serious scientist began their explanation with a warning about sensors and perception.

He interviewed radar operators who swore they'd tracked craft accelerating from standstill to thousands of miles per hour. He compared their stories with the explanations of atmospheric physicists who patiently described ducting, the way layers of warm and cold air could bend radar beams and create phantom objects that appeared to dart across the horizon.

At one briefing, a meteorologist showed him simulation data on "super-refraction." When temperature inversions trap radio waves, targets appear where no target exists. "Our instruments lie," she said, "but only because we taught them to see what we expect."

James filled pages with notes. Radar could extend the truth but infrared could twist it. Optical cameras, reliable to the naked eye, could be fooled by glare, parallax, and the shimmer of heat rising from the sea. Each device carried its own ghosts.

Still, some cases refused to flatten into errors. Multi-sensor detections involving radar, infrared, and visual confirmation all at once lingered in his mind like unresolved equations. The problem, James concluded, wasn't that the data were false, but that they were incomplete.

The Physics of the Impossible

At the Naval Research Laboratory, James was shown data from the now-famous "Tic Tac" encounter of 2004. On radar, the object appeared to drop from sixty thousand feet to sea level in less than two seconds. Pilots described a smooth white craft, featureless, accelerating without visible exhaust.

James compared those numbers to the equations of motion scratched on the whiteboard beside him. The accelerations implied would flatten any known aircraft and vaporize its hull from frictional heating. “If the numbers are correct,” the physicist beside him said, “then it isn’t aerodynamics. It’s something that looks a lot like magic.”

But magic, James knew, was not a concept used in physics. He began checking the provenance of the data, examining how the readings had been logged, what filters had been applied, and whether the sensors had been properly calibrated. A small misalignment in a radar dish could exaggerate velocity by orders of magnitude. A parallax error between two aircraft could make a slow-moving drone look like a dart of light.

The more he traced the chain of custody, the more he saw how extraordinary performance could emerge from ordinary noise. Yet in rare cases, even after every correction and verification, something remained, a residue that could not be explained by hardware faults or human error.

That remainder kept scientists honest, and kept James fascinated.

Artifacts in the Machine

In Colorado Springs, an aerospace contractor demonstrated how radar “multipath” reflections worked. By bouncing microwaves off a mirrored floor, the technician produced twin echoes that mimicked two targets performing synchronized maneuvers.

“Imagine this happening over the ocean,” he said. “You’d swear you had two objects dancing, but it’s just your signal coming home twice.”

James nodded, thinking of the Pacific fleet encounters. He began compiling what he half-jokingly called his *Field Guide to Illusions of the Sky*.

- Atmospheric ducting - creates phantom altitude shifts.
- Multipath - duplicates targets.
- Optical glint - makes a seagull look like a missile.
- Frame-rate aliasing - turns slow motion into instant acceleration.

He filled notebooks with these entries, the taxonomy of error. The deeper he went, the more he appreciated how easily the sky could deceive. A lens slightly out of focus could turn stars into disks. Heat shimmer could make lights twist and dance. Even the human eye, evolved for predators and sunsets, was poorly equipped for judging speed against an empty sky.

"Perception and instrumentation are partners in crime," one psychologist told him.

Materials and Myths

James's next stop was a materials lab in Nevada, where a physicist showed him a wafer of copper etched with microscopic spirals. "Metamaterial," she said proudly. When microwaves passed through it, they bent backward, refracting opposite to ordinary matter.

Scientists imagined that these designs might one day lead to cloaking devices, but for now the effect was small and fragile, useful for lenses but not for starships. "Anyone claiming broad-spectrum invisibility," she said, "is selling fiction."

Still, James couldn't resist visiting one more lab, where fragments allegedly "recovered from anomalous craft" were stored in sealed containers. The technician explained the process of isotopic analysis, describing how unusual ratios of magnesium or bismuth could, in theory, point to non-terrestrial origins.

The samples, when tested, proved disappointingly normal. Slight impurities, common in industrial alloys, had been mistaken for alien metallurgy. "Chain of custody's everything," the technician sighed. "Without it, science becomes rumor."

James left the facility reminded that wonder needed discipline. Without controls, the extraordinary collapsed into noise again.

Science of the Fleeting

The hardest lesson came when he tried to design a proper experiment. UAP didn't announce their arrival. They appeared where they pleased. "You can't schedule anomalies," a seismologist joked during an interdisciplinary workshop.

But there were precedents. Astronomers tracked gamma-ray bursts by networking telescopes around the world. Climate scientists studied extreme weather through statistical models of rare events. "Treat the sky like a distributed sensor," one data scientist suggested. "If it's real, more than one instrument will see it."

James pictured a network of ground stations, cameras, and radars all feeding into a shared cloud database, an astronomy focused on near-Earth space rather than the distant cosmos. Verification, replication, and transparency was the creed he saw forming.

He wrote in his field journal: *The first law of anomalies is to believe nothing until it occurs twice.*

The Human Equation

In a flight simulator at a training base, James experienced the illusions firsthand. The instructor dimmed the lights. A single point appeared on the horizon. After a minute, it seemed to drift and James felt compelled to track it. "That's autokinesis," the instructor said. "Your brain invents motion when the eye is bored."

Later they demonstrated the principle of size-distance inversion using two drones. One was small and close and the other large and distant, each moving at a different speed. Without range data, James guessed wrong every time.

Pilots, he realized, were not unreliable witnesses. They were human instruments operating at the edge of human capability. Stress, expectation, and fatigue each distorted perception in the same way that gravity bends light. And when radar or infrared echoed those distortions, the illusion hardened into evidence.

"People think seeing is believing," he told a colleague afterward. "In this field," she said, "seeing is nothing more than data acquisition, and every observation comes with error bars."

Building a Science of Anomalies

Months later, James presented at a conference of aerospace engineers and atmospheric scientists. His slides were unglamorous. They outlined sensor-fusion algorithms, propagation models, and chain-of-custody protocols. Yet the room was full.

He proposed a new research model that would deploy portable instrument pods, small autonomous units equipped with radar, infrared, optical, and magnetic sensors, to areas where sightings repeatedly occurred. Standardize the data schema so that civilian, military, and academic networks

could speak the same technical language. Share metadata openly, even if raw imagery remained classified.
A young engineer asked whether such rigor might drain the romance from the subject. James smiled. “Romance comes from discovery,” he said, “and discovery requires rules.”

Later that evening, walking through the conference courtyard, he thought of Kuhn’s line. *All great discoveries begin as anomalies.* He concluded that the real challenge was not locating anomalies, since there were plenty of those, but building a system disciplined enough to distinguish a true discovery from a mirage.

The Sky as Laboratory

One clear night in New Mexico, James stood beside a newly installed sensor pod, its antenna rotating silently under the stars. Data streamed in from every sensor. Thermal readings, radar pings, and optical frames, all carefully timestamped, cross-referenced, and stored for later analysis.

The desert air was still. A distant satellite traced its arc overhead, steady and lawful. Somewhere beyond that order, perhaps, lay something genuinely new.

For now, though, James was content to let the instruments listen. The hum of the equipment sounded almost like breathing, a quiet reminder that science itself was alive, always striving to separate wonder from error.

THE MACHINES THAT WATCH THE SKY

"Data is not information until it is interpreted."

- Daniel Bell -

The Room of Screens

The control room glowed with blue light. Monitors lined the walls like portals, each showing a different slice of the sky. On one screen, a satellite's thermal feed shimmered with ghostly heat signatures. On another screen, a destroyer's radar display pulsed with steady concentric circles. A bank of analysts murmured to each other in code and acronyms.

James La Pierre stood at the back of the room, notebook in hand, watching chaos turn into coherence. "This," a systems engineer told him, "is the place where stories turn into data, and sometimes, where data turns back into legend."

It was his first visit to a joint operations center, an unmarked building on the Virginia coast where naval, air, and intelligence teams worked to synchronize sensors that rarely spoke the same language. UAP investigation, he realized, wasn't a matter of finding flying saucers. It was about making sense of scattered signals that refused to align.

The Babel of Sensors

In his early research, James had believed that the problem of studying UAP was mainly philosophical, a matter of how belief and skepticism became entangled when confronting the unknown. But standing amid those screens, he saw that the problem was also logistical. Every device, whether a satellite, fighter jet, or destroyer, spoke its own distinct dialect of data.

Satellites saw wide but shallow. Enormous swaths of Earth in fine color gradients, refreshed every few minutes. Jets saw narrow but sharp. Bursts of electro-optical and infrared footage from high-speed chases. Ships saw patiently. Radar reflections rolling in with the rhythm of the sea.

Each system was both brilliant and blind. The satellite lacked temporal resolution. The jet's sensors could be fooled by clouds or glare. The ship's radar lost coverage when waves interfered. Only by fusing them could analysts begin to triangulate truth.

James watched two analysts align a radar return from a destroyer with infrared footage from an F-18. The radar displayed a blip, and the camera showed a dot of light. When they synchronized the timestamps down to the millisecond, the signals merged and one confirmed the other. For a moment, the unknown seemed to take shape.

"It's like assembling a mosaic," the engineer said. "Every tile is distorted. But get enough of them, and the picture emerges."

The Mathematics of Uncertainty

Later, in a quiet office, James sat with Dr. Aparna Chakraborty, a data scientist who specialized in aerospace sensor fusion. She drew diagrams on a tablet of intersecting ellipses that represented probability zones. "No single instrument tells you where something is," she said. "It only tells you where it might be."

She showed him the formulas for Bayesian inference and Kalman filters, mathematical frameworks designed to weave uncertainty into a coherent narrative. “The trick,” she explained, “is to let every system be wrong in its own way. Their mistakes cancel each other out.”

James asked whether this method ever produced certainty. Chakraborty smiled faintly. “Certainty is for textbooks. What we build here is confidence.”

He understood then why UAP analysis felt perpetually inconclusive. It wasn’t because scientists lacked imagination, but because the data itself was fractured. The sky’s complexity exceeded the architecture built to monitor it.

Barriers in the Code

Back in Washington, James met with a liaison from the Defense Innovation Unit. She described the institutional barriers that frustrated collaboration. Incompatible data formats, classified software, and international firewalls made global fusion almost impossible.

“Every contractor wants to protect their code,” she said. “Every nation wants to protect its airspace. And every sensor hoards its secrets.”

James pictured the sky as an orchestra with each instrument playing a different score. The melody of truth was there, but the notes refused to synchronize.

To address this challenge, researchers were developing secure middleware, platforms designed to translate between systems while keeping sensitive data encrypted. Some proposed tiered access. The military would see the raw stream, while scientists could analyze anonymized versions stripped of coordinates or unit identifiers. It was a compromise between secrecy and science, a bridge between classified and civilian worlds.

But the bridge was fragile. "The data is there," the liaison said. "What's missing is trust."

Machines That Watch the Sky

The next stage of James's exploration took him into the realm of artificial intelligence. At an aerospace firm in Palo Alto, engineers were training algorithms to recognize anomalies in radar and optical data, in effect teaching machines how to see the strange.

Rows of servers hummed as terabytes of imagery poured through convolutional neural networks. "We feed it everything," said Dr. Renata Garcia, one of the developers. "Drones, birds, jets, lightning. The model learns what's normal so it can flag what's not."

On the screen before them, the AI highlighted an object zigzagging across the frame. "That's an error," Garcia said, smiling. "A glitch in the sensor feed," the engineer said. "But that's the point. If it can learn to ignore the noise, maybe it can start to recognize what's real."

James asked whether the AI could be trusted. "Not yet," she admitted. "AI sees patterns, but it doesn't understand context. That's why we still need people in the loop."

He wrote in his notebook: *AI can recognize shapes, not meanings.*

At another lab, he met a cybersecurity expert who warned that AI systems could themselves be deceived. "A little adversarial noise, just a few extra pixels, and you can trick the algorithm into classifying an airplane as a cloud," she explained. "Now imagine what an adversary could do with that."

The lesson was clear. The future of detection depended not only on smarter machines, but on human oversight capable of questioning the machines' confidence.

The Human Factor

James's next stop was less digital and more human, the FAA's data analysis center in Oklahoma City. Here, veteran air-traffic controllers monitored reports from thousands of civilian flights.

He learned that commercial pilots had no standardized protocol for reporting UAP encounters. Most filed voluntary notes in safety systems meant for turbulence or bird strikes. "We're not supposed to say 'UFO,'" one pilot told him quietly. "It gets you labeled."

Military pilots had the opposite problem. They *had* to report, but their reports vanished into classified archives inaccessible to researchers. "Two worlds," an FAA official said. "One afraid to talk, one forbidden to."

James began outlining what he called a universal incident schema, a straightforward framework with fields for time, location, altitude, weather, sensor type, and witness account. It was designed to make civilian and military data interoperable while still protecting sensitive information.

He knew such standardization would require not only technical design but cultural change. "Destigmatization is a data-management problem," he later wrote. "If no one reports the anomaly, the system never learns."

When Data Becomes Story

One evening, James was invited to observe a simulation exercise at a defense research center. Analysts replayed a real incident from 2021. An unidentified object tracked by multiple sensors over the Atlantic.

On the first pass, the radar showed erratic movement. On the second, the infrared camera registered nothing. On the third, the satellite feed revealed a passing weather balloon caught in turbulent winds.

What fascinated James was not the solution but the process itself. Every analyst brought a different dataset and a different bias. The radar specialist insisted on electronic interference. The atmospheric physicist argued for ducting. The intelligence officer hinted at foreign surveillance drones.

By the end of the session, consensus formed not around certainty but probability with *95 percent mundane* and *5 percent unknown*. The remaining 5 percent was what kept everyone coming back.

“Data doesn’t solve mysteries,” one analyst told him as they packed up. “It just narrows them.”

Toward a Common Sky

As months passed, James watched momentum building around the idea of unified reporting. In Congress, legislators debated the creation of a public portal modeled after France’s GEIPAN, a system that would allow pilots, civilians, and military personnel to submit encounter reports in a standardized format, with sensitive data carefully redacted.

Internationally, the European Space Agency and NASA began discussing metadata standards for anomaly tracking, treating

UAP not as curiosities but as potential safety issues. The movement felt pragmatic, not sensational.

When James spoke at a symposium in Brussels, he argued that harmonized reporting would benefit everyone, from scientists seeking reproducible data to defense planners working to protect national airspace. “It’s not about believing or disbelieving,” he said. “It’s about cataloging what’s in our skies with the same discipline we apply to what’s under our microscopes.”

The applause that followed was polite but sustained. In that moment, James sensed that the stigma surrounding the subject was finally beginning to erode.

The Sky Network

At dawn on his final field assignment, James boarded a research vessel in the Pacific equipped with the latest sensor suite. On board were radar, optical telescopes, infrared cameras, and AI-assisted tracking software. The ship’s captain explained that they were participating in a multinational experiment that linked satellites, aircraft, and sea platforms into a single synchronized network, a living laboratory of the atmosphere.

As the sun rose, the instruments began their slow rotation. Every minute, data pulsed through fiber links into cloud servers thousands of miles away. “If anything unusual flies over us today, it won’t belong to any one nation’s secret,” the captain said.

James watched the horizon, where sea met sky in seamless gray. Somewhere within that vastness, perhaps, another mystery waited, a signal too faint or a pattern too rare to be noticed yet. But for the first time, he felt that the world was learning how to listen together.

Information from the Noise

Months later, back in his office in D.C., James sifted through the compiled data from that mission. The files filled terabytes of storage, billions of data points distilled down to only a few scattered anomalies. Yet what mattered wasn't whether one of them turned out to be extraordinary.

What mattered was that they existed within a coherent system, time-stamped, verified, and replicable. The era of fragmented anecdotes was coming to an end.

He looked out the window toward the contrails above the Potomac. Airliners carved their steady arcs across the morning sky, ordinary and magnificent.

In the margins of his notebook, he wrote: *Data becomes knowledge only when shared.*

And for the first time, he believed that humanity might yet learn to interpret its own sky.

THE SEARCH BEYOND EARTH

"The universe is not only stranger than we imagine. It is stranger than we can imagine."

– J.B.S. Haldane –

The Conference Room at Ames

The air in the NASA Ames auditorium felt thin, filtered, like it had passed through too many ducts. A mural of Europa's icy surface stretched across the back wall, and the faint hum of air-conditioning mixed with the sound of scientists adjusting microphones. James La Pierre sat near the back, notebook open, watching as one of the agency's planetary chemists advanced to the podium.

The slide behind her read *"Biosignature Ambiguity in Exoplanetary Atmospheres."*

It was not the sort of title that made headlines, but the tension in the room was unmistakable. Here were the people who had spent their careers calibrating instruments so precise that they could read a molecule's spectral fingerprint across fifty light-years, yet even they could not agree on what truly counted as evidence for life.

James had come to Ames to trace the thin and often controversial boundary where astrobiology met the emerging UAP debate. In one corner of the field, scientists were refining algorithms to detect methane on distant planets. In another, military analysts were debating objects moving through

Earth's own skies. Both, in their own way, were chasing anomalies.

A physicist from MIT took the stage next. "When we talk about biosignatures," she began, "we have to separate what is truly biological from what is simply interesting. Ozone can be produced by photosynthesis, but it can also form through the photolysis of carbon dioxide under the right kind of stellar spectrum."

A murmur rippled through the audience. The point was simple but devastating. Even if a telescope found oxygen on another world, it might not mean life. The universe, it seemed, was a trickster.

James jotted down a note: *Extraordinary claims must pass through ordinary methods.*

He'd heard the phrase before, attributed to one of the astrophysicists working on the Webb telescope, but it summed up what he was seeing here. Astrobiology, like the study of UAP, had become a lesson in humility. Both fields wrestled with the same riddle. How to extract truth from data that resisted easy interpretation.

The Scientific Frame

Later that afternoon, over burnt coffee in the Ames cafeteria, James met with Dr. Riya Shah, a spectral analyst from the SETI Institute. She was young, sharp, and weary from years of defending her work against sensationalism.

"People assume we're looking for little green men," she said, glancing toward a group of journalists gathering outside the glass doors. "What we're actually searching for is imbalance, atmospheres that defy thermodynamic expectations."

James leaned forward. "And that imbalance could mean life?"

"Or it could mean chemistry we don't understand yet," Shah replied. "The universe is good at producing false positives."

She opened her laptop and displayed a series of color-coded graphs showing oxygen bands, methane dips, and carbon dioxide spikes. "Each of these lines," she said, "is a whisper from a planet. The real challenge is figuring out which ones are telling the truth."

He thought about that as she spoke. Lies hidden in data were something he understood well from his earlier reporting on defense radar systems. Misleading readings were common. Layers of atmospheric ducting, multipath reflections, and even swarms of insects could create echoes that appeared extraordinary.

Shah must have read his expression. "You see the parallel, don't you?"

"Yes," James said slowly. "Same problem, different scale."

"Exactly," she said. "In both cases we are fighting against many kinds of noise, from technical to environmental to psychological. The cosmos never offers simple answers. You have to earn them."

That line stayed with him long after the meeting ended.

A Universe of False Positives

Over the following months, James traveled between institutions like Caltech, Harvard's Center for Astrophysics, and the European Space Agency in Paris, following the trail of research that blurred the boundary between astrobiology and UAP investigation.

Everywhere he went, he found scientists engaged in a quiet war against misinterpretation.

At the European Space Operations Centre in Darmstadt, an atmospheric physicist named Erik Neumann guided him through recent findings on transient luminous events including sprites, blue jets, and gigantic jets that flickered above thunderstorms near the edge of space.

“These were considered myths until the 1990s,” Neumann said, pointing to footage on a monitor. The images showed tendrils of blue light branching upward like jellyfish. “Now we can measure their temperature, their electron density, and their altitude. They’re beautiful, but they also account for hundreds of so-called ‘light anomalies’ reported by pilots.”

Neumann turned to James. “The atmosphere is an artist of deception. Before we imagine visitors from another world, we should learn what our own planet can paint in the sky.”

James smiled at the phrasing. It reminded him of the UAP analysts he’d interviewed at the Pentagon. They spoke in calm, empirical, and almost poetic tones. The closer one got to the unknown, it seemed, the quieter the language became.

He wrote in his notebook that night: *Science doesn’t kill mystery. It teaches it discipline.*

Technosignatures and the New SETI

In California, James met Dr. Harold Lin, one of the principal researchers in the *Breakthrough Listen* project. Lin’s office was cluttered with signal charts and mug stains. He spoke quickly, toggling between radio spectra and infrared graphs.

"SETI used to mean one thing," Lin said. "We listened for radio beacons. But civilizations, if they exist, might not use radio at all. They might leave waste heat, or industrial pollutants, or orbital artifacts we can detect optically."

He paused and brought up an image of a star field. "This," he said, "is the new SETI, technosignature astronomy. We are searching for the fingerprints of industry rather than the traces of intention."

James studied the screen. "And what about UAP data?" he asked. "Does any of it interest you?"

Lin leaned back. "Interest, yes. Reliability, no. The problem is that UAP data isn't public, reproducible, or standardized. If you can't share the raw data, you can't do science."

James nodded, remembering the classified briefings he had attended months earlier. Defense systems had captured anomalies, objects moving faster than any known aircraft, but the data remained locked behind clearance levels that few scientists could access.

Lin gestured toward his monitors. "The day we can apply our tools including spectroscopy, calibration, and time correlation to a verified UAP dataset, we will finally know whether we are chasing physics or fiction. Until that day, they are only stories."

The line wasn't dismissive. It was procedural. Lin wasn't a skeptic or a believer. He was a scientist defending the integrity of his toolkit.

The Fermi Question

One evening, back at his apartment in Washington, D.C., James reread his notes from the past six months and realized that nearly every scientist, regardless of field, had referenced the same haunting question.

"Where is everybody?"

The Fermi Paradox, posed in 1950, hung over every conversation like a ghost. Billions of stars, billions of planets, and yet no clear evidence of another civilization. For every argument suggesting the galaxy should be teeming with life, there was an equally convincing explanation for its silence.

Some believed that intelligent species eventually destroyed themselves. Others suggested that advanced civilizations chose to remain hidden. Still others argued that human perception, limited by its own bandwidth and methods, was simply too narrow to detect them.

James wrote: *Absence of evidence is not evidence of absence, but it's a stubborn silence, nonetheless.*

He imagined the scientists at Ames poring over spectra from exoplanet atmospheres, searching for traces of meaning hidden in the static. In a way, they were doing what humanity had always done, listening for echoes in the dark.

Biology Beyond Earth

At a lab in Cambridge, he met with Dr. Elena Bains, a chemist studying alternative biochemistries. Her office was lined with flasks of viscous, colored liquids. She greeted him with a handshake that smelled faintly of sulfur.

"We've been water chauvinists for too long," she said with a grin. "Life might not always look like us. It might not even *feel* like us."

She described her experiments using ammonia and sulfuric acid as possible solvents for alien biochemistry. In sealed laboratory chambers, her team simulated the conditions found on Titan and Venus. Some of the mixtures produced self-organizing droplets that behaved like primitive analogues of living cells.

"Are they alive?" James asked.

"Not by our definition," she replied. "But they hint that life could be chemistry finding new ways to persist."

He asked whether such forms could ever develop intelligence. She paused for a moment and said, "That is the harder leap. But Earth has already shown that intelligence can evolve more than once. Octopuses, crows, and dolphins all prove the point. Maybe the cosmos knows how to repeat the trick."

James left the lab feeling unexpectedly hopeful. For decades, the search for life had been viewed as a speculative pursuit, yet here, in a basement filled with beakers and gas lines, the idea seemed practical, almost routine.

The Threshold of Proof

Months later, James stood on the observation deck of the Keck Observatory in Hawaii, watching technicians calibrate the telescopes. The wind carried volcanic dust. Somewhere in the data streams below, photons from distant worlds were being transformed into numbers, and those numbers into arguments about life.

He thought about what it would take for a UAP claim to meet the same standard of proof. Multiple, calibrated detections. Independent verification. Reproducibility. The same thresholds demanded for exoplanet biosignatures or gravitational waves.

He realized that the scientific bar wasn't set high to discourage belief. It was there to protect discovery from collapse. Without rigor, wonder turns to noise.

In his field notes he wrote: *The extraordinary must survive the ordinary tests.*

That night, as the telescope dome rotated above him, he thought about the scientists he had met. He remembered their exhaustion, their patience, and their quiet respect for uncertainty. They were not chasing aliens. They were building frameworks strong enough to hold the weight of whatever truth might one day appear.

The Convergence

By the time James returned to Washington, the landscape had shifted again. Congressional hearings on UAP had entered the public domain. NASA announced the creation of a panel to study unexplained aerial phenomena with scientific transparency. And within the astrobiology community, discussions were turning toward collaboration rather than avoidance.

At a press briefing, a planetary scientist summarized it clearly, "If we ever confirm a truly anomalous object within Earth's atmosphere, it will happen because we used the same tools we rely on for studying exoplanets. The tools of physics, chemistry, and reproducibility rather than belief."

James closed his notebook for the last time that evening. The worlds of UAP research and astrobiology were no longer separate curiosities. They were beginning to converge, each one forcing the other to evolve.

He realized that in both realms, the goal was the same. Not to prove the extraordinary, but to understand the possible.

And that, he thought, might be the beginning of a new kind of science.

Diagnosing the Sky

The rain in Washington fell in relentless sheets that week, and James found himself once again inside the sterile quiet of a Pentagon annex, an office that smelled faintly of paper and ozone. A digital projector cast radar traces across a frosted glass wall. He was there to meet Colonel David Singh, an Air Force analyst who had become one of the more reluctant members of the government's newly formed UAP working group.

"Everyone wants this to be the moment," Singh said, rubbing his temples while a technician cycled through the radar data. "But the truth is messy. Ninety-eight percent of what crosses our screens has an explanation like ducting, ionospheric bounce, miscalibration, or atmospheric clutter. The other two percent is not mystery. It is missing context. Incomplete data pretending to be profound."

James watched as the colonel froze one frame, a streak of light cutting sharply across the grid. "Looks dramatic," Singh said. "But if you overlay the weather radar, it turns out to be a transient luminous event, an electrical discharge high in the atmosphere."

"Sprites," James said, remembering his conversation with Neumann in Darmstadt.

"Exactly. They fooled us for decades." Singh exhaled. "This is what I tell my team. Nature is a better engineer than we are."

The colonel leaned closer and spoke quietly. "But sometimes the data refuses to fit anything we know. That's when I start questioning the instruments themselves, from digital compression to telemetry loss to rolling shutter effects. Until every technical flaw is ruled out, we have no right to use words like extraterrestrial."

James nodded. The colonel wasn't dismissive. He was disciplined. He was applying the same process an astrobiologist would to a suspected biosignature. Rule out the local before invoking the cosmic.

In his notes, James wrote: *Classification is not just about secrecy. It's about uncertainty.*

The Orbit of Artifacts

A month later, James was at the European Southern Observatory's control center in Chile, where the sky at night felt closer than skin. He had come to understand how the growing constellation of human satellites complicated even the most careful observations.

A young orbital analyst named Sofia Bassa guided him through a map filled with moving dots, thousands of satellites tracing paths around the Earth. "This," she said, "is our new night sky. Every one of these tracks can look like an anomaly if your instruments are even slightly out of calibration."

On her screen, bright streaks crossed the telescope images, sunlight flashing off metal panels hundreds of miles above Earth. "We can predict these now down to the second," Bassa explained. "But in defense or civilian aviation data, without accurate orbital catalogs, one reflected glint can easily look like something moving with intent."

She zoomed in on one frame. "See that flicker? It's a Starlink flare. That's what thousands of reports look like."

James watched the data scroll by, thinking of the irony. Humanity had filled the sky with its own ghosts. Our search for the other had been haunted by our own inventions.

Sofia folded her arms. "It's not that we're debunking everything," she said. "We're just cleaning the lens."

The Laboratory of Life

In a sterile lab in Pasadena, James stood behind a glass partition while Dr. Matthew Haqq-Misra adjusted a reactor chamber. Inside, a mixture of gases swirled under ultraviolet light. "This is a miniature Venus," Haqq-Misra explained. "We are testing whether photochemistry alone can produce false biosignatures, oxygen and ozone without life."

He flicked a switch, and the gas mix shimmered faintly purple. "You'd be amazed how many 'life-like' signals we can fake."

James took notes quickly. The experiment echoed a lesson familiar to astrobiologists. Extraordinary signals often turned out to have ordinary explanations. Yet beneath the careful skepticism he sensed something else, a quiet kind of optimism that came even from disproving the extraordinary.

"What about methane?" James asked. "It's been showing up on Mars again."

Haqq-Misra smiled. "Methane is the universe's favorite prank," he said. "It can come from life, from rocks, or from simple instrument error. The real challenge is correlation, figuring out whether those methane spikes align with seasons or with geological activity."

He turned off the chamber. “Astrobiology isn’t about finding aliens,” he said. “It’s about making sure that when we do find something, we know it’s real.”

That line echoed what Lin had said at SETI, what Shah had said at Ames, what Singh had implied in the Pentagon basement. Every field was converging toward the same ethic of rigorous humility.

The Technosignature Frontier

James’s next stop was the Berkeley SETI Research Center, where a wall of servers filled the room with a low electric hum. Graduate students sat quietly at their terminals. Their eyes fixed on scrolling lines of code that pulsed across black screens. They were searching for technosignatures, the faint fingerprints of technology woven into the endless static of the cosmos.

Dr. Jill Wright, a senior researcher, pointed to a scatter plot speckled with faint red clusters. “These are the outliers,” she explained. “Signals that don’t match any known astrophysical source. Most turn out to be interference from Earth, but every so often one appears that makes us stop and look again.”

She pointed to one labeled *Candidate 873-2*. “Narrowband emission, repeating every 22 hours. It turned out to be a weather satellite. But for a week, half this building couldn’t sleep.”

James smiled. “So the thrill is still there.”

“Always,” Wright said. “But we’ve learned to trust the process, not the excitement.”

She leaned back and crossed her arms. “Here’s the thing. If someone ever brings us credible UAP data that is properly timestamped, verified by multiple sensors, and cross-calibrated, we would analyze it just like any other dataset. This is not about belief. It’s about the hierarchy of evidence.”

James understood. In SETI, as in UAP research, the tools were the same and only the culture was different.

The Fermi Paradox Revisited

In Geneva, James attended a symposium devoted to “The Great Filter,” the theoretical barrier that might explain why so few civilizations reach the point of interstellar contact. The hall was filled with philosophers, biologists, and astrophysicists debating in precise, statistical language the many ways a species might end before it ever learns to cross the stars.

A slide projected onto the wall read simply, “Where is everybody?”

During the coffee break, James spoke with Dr. Anders Grinspoon, who had written about planetary intelligence as an emergent ecological process. “Maybe the reason we haven’t heard from anyone,” Grinspoon said, “is that civilizations that survive learn not to broadcast.”

James raised an eyebrow. “Because of security?”

“Because of maturity,” Grinspoon replied. “Survival may mean blending into the noise.”

That idea stayed with him. Perhaps the absence of evidence was not an emptiness but a kind of boundary. Civilizations, like individuals, might eventually move toward silence, seeking balance instead of conquest.

When the session resumed, a young researcher presented models suggesting that the probability of two intelligent civilizations coexisting and communicating was statistically minuscule. James found himself both comforted and unnerved. The math, like the cosmos itself, was indifferent.

He wrote later: *The silence may be the message.*

The Ocean Worlds

From Geneva, James traveled to the Jet Propulsion Laboratory in California, where engineers were preparing instruments for the *Europa Clipper* mission. In the cleanroom, Europa's icy terrain was simulated on large, illuminated panels.

Dr. Ana Pappalardo explained how radar sounders would probe beneath the moon's frozen crust. "We know there is an ocean down there," she said. "What we do not yet know is whether it contains the ingredients for life including energy gradients, organic compounds, and enough time for complex chemistry to emerge."

James asked, "And if it does?"

"Then we'll finally know that life is not a miracle. It's a consequence."

Her words had the gravity of revelation. The team spoke of ice fractures and magnetic fields, but what they were really chasing was a cosmic mirror. Every probe and every spectrum was a question about ourselves. *Are we an accident or a pattern?*

Later, as he stood in the parking lot watching the sun dip behind the San Gabriel Mountains, James realized that the search for life beyond Earth had never been about proof alone. It was about identity.

The Threshold of Evidence

By autumn, James had filled half a dozen notebooks. Each page returned to the same ideas of calibration, reproducibility, and context. Whether he was studying UAP or exoplanet atmospheres, the standard of proof never changed. It required multiple, independent detections that were consistent across instruments and in agreement with the laws of physics.

He remembered Lin's words, "If it's real, it will survive cross-validation."

Back in D.C., he sat in on a closed NASA session where analysts discussed a new data-sharing proposal between AARO and SETI researchers. The idea was simple but revolutionary. Allow sanitized, declassified UAP data to be studied by scientists using the same frameworks as astrobiology.

A senior official concluded the meeting with a single sentence that hung in the air like a verdict. "We owe it to the public to be skeptical and sincere at the same time."

For the first time, James sensed that the two worlds of defense and science were beginning to speak the same language.

Reflections on Intelligence

That winter, James met Dr. Valentina Roth, a cognitive biologist studying convergent evolution. She showed him videos of octopuses unscrewing jars, crows bending wires into hooks, dolphins using marine sponges as tools.

"Intelligence," she said, "isn't rare. It's opportunistic. It evolves wherever it helps something survive."

She tapped a model of a neuron on her desk. "The real rarity is not thought," she said. "It is technology. To make tools, you need hands, fire, and shared learning. You need cooperation."

James remembered Grinspoon's remark about maturity. Perhaps intelligence was not only about invention but also about restraint, the ability to recognize and respect one's limits.

That night he wrote: *The universe might be full of minds, but not all of them need to build machines to matter.*

The Discipline of Wonder

A year after that first conference at Ames, James returned for another symposium. The same mural of Europa glowed against the wall, but the tone in the room had changed. There was less skepticism now and far more structure.

During a panel on UAP transparency, one speaker said, "Our best safeguard against fantasy is method."

James closed his notebook. In the margin of his first page, he'd written the same phrase twelve months earlier. The circle was complete.

Outside, the night was clear. A faint satellite crossed the stars above the parking lot, a line of reflected sunlight that was perfectly predictable and unmistakably human.

James knew that most mysteries would fade under careful examination, yet some might endure.

Science, he realized, was never the destroyer of mystery but its caretaker.

For the first time in years, he felt that the search for life, truth, and understanding was no longer split between Earth and the heavens. It had become one continuous horizon.

PART III

THE MIND BEHIND THE PHENOMENON

**"We do not see things as they are,
we see them as we are."**

- Anaïs Nin -

Part III follows James La Pierre into the social and psychological core of the UAP mystery, where the questions shift from radar returns to human perception and belief. The focus moves from instruments to minds, beginning with the limits of vision, cognition, and bias that shape what witnesses think they have seen. It then enters the uncertain terrain of memory reconstruction, sleep paralysis, and belief formation, where experience and interpretation blur.

These chapters show how private encounters become public narratives. Media, mass sightings, and digital culture transform fleeting perceptions into enduring myths that blend science, folklore, and authority. Alongside this cultural spread, James examines academia itself, where stigma and curiosity collide as scholars attempt to study UAP without losing credibility. Part III reveals that UAP are not only unexplained events but reflections of how humans assign meaning. In this sense, the mystery becomes a mirror, exposing belief, power, and the persistent human need to understand what lies just beyond certainty.

THE SKY WITHIN US

"The human mind is the greatest filter of reality."

- William James -

The Minds Behind the Lights

The conference room smelled faintly of coffee and carpet cleaner. James La Pierre sat near the back, notebook in hand, watching as Dr. Karen French, a cognitive psychologist from the University of Sussex, advanced to her next slide. On the screen glowed the words *"Perception Under Ambiguity."*

French gestured to an image of a faint light over a horizon. "When the brain encounters uncertainty," she said, "it fills in the blanks with stories."

James looked around the room. Pilots, neuroscientists, and sociologists shared the table. Some wore flight uniforms while others wore tweed jackets. All were here to answer a single question. *Why do so many people see extraordinary things in the sky?*

French continued. "Night vision degrades color, erases depth, and exaggerates motion. A star twinkles and we imagine movement. A drone tilts and we imagine control. The more frightened or expectant we are, the more certain the illusion becomes."

James scribbled in his notes: *The sky as mirror, not mystery.*

The Fragile Senses

A week later, he visited a U.S. Air Force training center in Nevada. Pilots were running night-flight simulations using infrared helmets. An instructor dimmed the lights and played back cockpit footage from an earlier exercise. A flickering sphere darted across the desert.

"That was Venus," the instructor said. "Altitude thirty thousand feet, azimuth two thirty-five."

The pilots laughed. James did not. He had seen the same clip on social media, stripped of context, labeled *'unknown aerial craft.'*

Later, the instructor explained the basics of misperception. At night, without fixed reference points, angular velocity deceives the eye, nearby lights appear to accelerate, and distant ones seem stationary. Stress, fatigue, and motion blur do the rest.

James thought of Dr. French's words. The human visual system, it seemed, was less an instrument of truth than a negotiation between light, expectation, and fear.

That evening he wrote in his notebook: *The unknown begins in the retina.*

Patterns in the Noise

Back in London, James met a psychologist who specialized in pareidolia, the human tendency to find meaning in random patterns. Her office walls were covered with photographs of clouds that resembled faces, birds, and even what looked like spacecraft.

"When the visual cortex searches for coherence," she said, "it rewards itself with recognition. The same mechanism that helps us find tigers in the grass makes us find saucers in the clouds."

She showed James an experiment. Participants stared at static until shapes emerged. Within minutes, half of them reported "objects." Some were certain they saw metallic disks. None existed.

"Perception," she added, "is hypothesis testing in real time. When the hypothesis is aliens, the test rarely fails."

James left the office uneasy. The data were persuasive, yet he could not shake the sense that statistics alone failed to explain the emotion he had heard in the voices of witnesses. Something deeper was at work, something rooted in memory, belief, and the need for belonging.

The Night Visitors

In Arizona, James met a man named Richard Elders who claimed to have been abducted. They spoke in a quiet diner outside Flagstaff. Richard's voice trembled as he described waking in bed unable to move, a bright light at the window, a presence in the room.

"I tried to scream but nothing came out," he said.

James recognized the pattern from the literature on sleep paralysis. The body frozen between REM and waking, the mind half-dreaming, half-aware. When Richard described floating above his bed, James heard the vocabulary of neurology filtered through the imagery of television.

Later, he met Dr. Amira Cicogna, a neurologist who had studied these experiences. She showed him brain-wave scans taken during episodes of REM intrusion. "They feel awake," she explained, "but their motor neurons are frozen.

The hallucinations borrow whatever symbols the culture provides, including spirits, demons, or aliens. The physiology is universal, but the story changes from one culture to another."

James asked what she told patients who insisted their experiences were real.

"I tell them both explanations can be true," she said. "Their fear was real. The experience changed them. The cause, however, was internal."

The Crowd and the Sky

That summer James attended a community meeting in rural Texas where dozens of residents claimed to have seen "dancing lights." The atmosphere was charged. Someone played cellphone footage on a projector of three dots moving against blackness.

As the crowd murmured, a retired sheriff stood and declared, "They're here again." Applause followed.

James noted how quickly individual uncertainty became collective conviction. When he later interviewed attendees separately, their descriptions converged on size, color, and formation, details that had never appeared in the original footage.

At a university lab, social psychologist Dr. Maria Giordano explained why. "Memory is social," she said. "Once a group agrees on a version of events, the mind edits itself to match the consensus. Conformity feels safer than doubt."

She showed him archival footage from classic conformity experiments. Participants insisted a shorter line was longer simply because others said so. "Now scale that up to a sky full of light," she added. "The result is folklore in real time." James realized that collective sightings were less about objects in the air than about trust on the ground.

The Weight of Stigma

Not everyone was willing to talk. Several commercial pilots had declined interviews outright, fearing mockery from colleagues. One former Navy aviator finally agreed to meet James in a quiet bar near Norfolk.

"We're told to report everything," the pilot said, "but report a UFO and you'll spend more time in evaluations than in the cockpit."

He described the uneasy balance between safety reporting and reputation. "You want the data logged, but you don't want to be the story."

James encountered the same silence in academia. A physicist at a major university confided, "If I submit a paper on UAP, reviewers will assume I've lost my mind."

Sociologist Dr. Hannah Wendt helped him see the pattern. "Stigma acts like a firewall," she said. "It keeps institutions safe from embarrassment but also from discovery. Boundary-work, we call it. It decides what counts as science."

James asked if that boundary could ever change.

"Only when the cost of ignorance exceeds the cost of curiosity," she replied.

Screens and Shadows

In a dim newsroom in New York, James watched as an editor replayed cockpit footage recently released by the Navy. The video had already gone viral online.

"Fifteen million views in twenty-four hours," the editor said. "Most of the comments think it's aliens."

James knew the algorithms favored outrage and wonder over nuance. Online, repetition became evidence and virality became verification.

He later spoke with digital-media researcher Dr. Michael Rand. "Social platforms reward engagement, not accuracy," Rand said. "The more uncertain a clip, the more it spreads. Our brains evolved for campfire stories, not compressed pixels."

Rand showed him data on the "illusory truth effect." Statements repeated online and even when flagged as dubious grew more believable over time. "The internet," he said, "is the largest psychological experiment in history, and no one's running the control group."

As James left the building, he glanced at a billboard advertising a science-fiction film. A glowing saucer hovered above a city skyline. Fiction and reportage had become indistinguishable décor.

Archetypes and Ancestry

To understand why the same images persisted for generations, James met comparative mythologist Dr. Jeff Kripal at Rice University. Kripal spread a stack of illustrated manuscripts across his desk showing winged beings, fiery wheels, and angels descending.

"These are from medieval Europe," he said. "If you change the vocabulary from halo to disk and angel to pilot, you get today's sightings."

He explained that the sky had always been a projection screen for human longing. "In one era, we call them gods. In another, extraterrestrials. The motif survives because it speaks to the same psychological need. Something greater is watching."

James felt the weight of centuries in those drawings. The continuity between ancient and modern sky visions was undeniable. The mystery, perhaps, was less about what people saw than why they needed to see it.

Minds in Conflict

Back in Washington, James reviewed the data he had gathered. Cognitive biases explained much, and cultural context explained even more. Yet a handful of reports remained, including multi-sensor detections, corroborated radar and infrared signatures, and credible witnesses.

He met with physicist Dr. Kevin Knuth, who had studied such cases. "Psychology explains most sightings," Knuth said, "but not all. Some data points persist after every mundane filter is applied."

James asked whether he believed in extraterrestrial visitation.

Knuth smiled faintly. "Belief is irrelevant. The question is whether our models can accommodate what we observe. If they can't, we revise the models."

James realized that skepticism and openness were not opposites but partners in progress. The mind could mislead, but it could also learn.

The Human Aftermath

At a small support meeting in Portland, James listened to experiencers share their stories. Some spoke of lights, and others of contact. What struck him was not the content but the emotion, including fear, wonder, and sometimes gratitude.

A woman named Elise described how her experience had given her purpose. “People think we’re crazy,” she said, “but finding each other saved me.”

Sociologists call this identity work. The transformation of an anomalous event into a source of belonging. Within these groups, marginalization becomes meaning.

James understood then that UAP were not merely phenomena to be measured but experiences to be lived. Whether caused by psychology, physics, or something uncharted, they reshaped lives.

The Mirror of the Unknown

Months later, James returned to the same conference room where his journey had begun. Dr. French was again at the podium. Her closing slide read, *“Perception is participation.”*

“The mind,” she said, “does not just record the world. It constructs it.”

As the audience dispersed, James lingered by the window, watching contrails fade into the afternoon sky. He thought of every witness, every skeptic, every believer he had met. The lights above might be stars, satellites, or illusions, yet the stories they inspired were undeniably real.

In the end, he concluded, UAP were a conversation between sky and psyche, technology and myth, evidence and emotion. The mystery was not simply out there. It was within the observer, refracted through memory, fear, and hope.

He closed his notebook. The search for the unknown, he realized, was as much about understanding humanity as it was about understanding the heavens.

THE CLASSROOM OF THE UNKNOWN

"Education is not the filling of a pail,
but the lighting of a fire."

- W.B. Yeats -

The Lecture That Wasn't Supposed to Happen

The announcement appeared quietly on a departmental bulletin board. *"Special Seminar: The Scientific Study of Anomalous Aerial Phenomena."* No university logo accompanied it, only a room number and time. When James La Pierre arrived that afternoon, half the seats were filled by graduate students, half by curious locals. A physics professor in a corduroy jacket stepped to the podium and adjusted the microphone.

"This talk," he began carefully, "is off the record."

He clicked to a slide showing a still frame from a Navy cockpit video. A glowing sphere darted across the infrared display. "The question," he continued, "is not whether you believe. The question is whether we, as scientists, are willing to measure."

That moment, James realized, captured the state of academia's relationship with Unidentified Aerial Phenomena. The subject was inching its way out of taboo, one cautious seminar at a time. Universities that once laughed at the notion of "flying saucers" were beginning to reconsider, pressed by new data, government reports, and student curiosity.

The Stigma That Wouldn't Die

James spent weeks traveling from campus to campus, asking faculty why the subject remained so radioactive. Many answered the same way. Fear. A professor at a prestigious American university admitted that even mentioning UAPs in a proposal could sink a grant. "Peer review is a social system," she told him. "If your peers think you've lost your scientific footing, they will quietly step away."

In an office lined with awards, Dr. Thomas Eghigian, a historian of science, explained the legacy of ridicule. "During the Cold War," he said, "intelligence agencies deliberately trivialized the topic. The idea was to prevent public panic and protect classified projects. The collateral damage was academic curiosity."

James saw the pattern clearly. Once a topic was branded pseudoscience, no amount of new data could rescue it without institutional permission. The stigma had become self-perpetuating, woven into the very language of scientific credibility.

The Boundary of Science

At a philosophy colloquium in Paris, James listened as Dr. Elena Gieryn lectured on "boundary work," explaining how scientists draw lines between legitimate and illegitimate inquiry. She argued that UAPs were excluded not because of the data, but because of the narrative surrounding them. "Science defines itself by what it refuses," she said. "The sky is not the problem. Reputation is."

Later, over espresso in the campus café, James asked her whether that boundary could ever shift. Gieryn smiled.

"Boundaries are negotiated. Once governments, funders, or technologies change the conversation, the line moves. It always has."

Her words echoed through his travels. Perhaps the UAP question was less about belief and more about institutional evolution.

The Bridge Builders

In Florida, James visited the small but ambitious headquarters of the Scientific Coalition for UAP Studies (SCU). Inside, researchers from aerospace, psychology, and data science worked quietly in front of glowing monitors. Their current project analyzed radar and infrared data from the USS Nimitz "Tic Tac" incident.

Dr. Kevin Knuth, a physicist and former NASA researcher, showed James a series of graphs. "We're applying the same standards we'd use for any aerospace anomaly," Knuth explained. "Signal-to-noise ratios, Doppler tracking, flight dynamics. Nothing mystical about it."

Nearby, another team from UAPx prepared field equipment, including optical telescopes, infrared cameras, and spectrum analyzers, for a coastal observation campaign. "We're not here to prove aliens," said team lead Matthew Powell. "We're here to collect clean data."

James admired their approach, combining citizen science with professional rigor. Both organizations embodied the shift he had noticed since the Pentagon's 2021 report, a cautious yet credible re-entry of UAP research into the realm of open inquiry.

The Universities Stir

At Stanford University, James met a medical researcher analyzing tiny metal fragments said to be connected to reported UAP events. The lab smelled of ozone and acetone. "We're not chasing myths," the scientist said, leaning over a mass spectrometer. "We're analyzing isotopic ratios. If something is unusual, we document it. If it's normal, we say so."

In Würzburg, Germany, an engineering professor showed James how his students used declassified UAP radar data to teach sensor reliability. "Every mystery teaches us calibration," he said. "It's the best way to learn humility with instruments."

At Arizona State University, James attended a lively seminar titled *"From SETI to UAP: Evidence and Imagination."* Students debated whether unidentified phenomena should be studied by astronomers, sociologists, or both. The professor concluded, "We're not studying aliens. We're studying ambiguity."

It struck James that academia was finally allowing itself to use UAP as a lens, focusing less on proving or disproving visitation and more on examining how knowledge itself is created.

The Publishing Wall

James's next stop was a small editorial office at a scientific journal. The editor, an astrophysicist with silver hair, sighed as he spoke. "We receive good UAP papers from time to time. They include solid math and careful analysis, but we cannot publish them. Reviewers revolt."

When James asked why, the editor replied, "Brand protection. A journal's prestige depends on its signal-to-noise ratio, and UAPs are still considered noise."

He showed James a rejected manuscript that had modeled flight dynamics from military radar data. It was as technical as any aeronautics study. "If this paper were about plasma instabilities," the editor said, "we'd run it tomorrow."

That conversation revealed to James the invisible economics of credibility. Journals, funders, and departments all guarded their brands. Until those guardians changed their risk calculus, UAP studies would remain intellectually homeless.

Seeds of a Discipline

At a science policy conference in Washington, James joined a roundtable of professors, defense officials, and nonprofit researchers. The discussion centered on how to bring UAP studies into the academic mainstream.

One proposal suggested interdisciplinary centers where physicists, psychologists, and sociologists could collaborate without stigma. Another urged NASA and the All-Domain Anomaly Resolution Office to fund open research grants. A third envisioned graduate programs combining aerospace engineering with science and technology studies.

As the session closed, a senior policy adviser remarked, "If astrobiology could rise from speculation to legitimacy once exoplanets were confirmed, so can this field. We just need the courage to formalize it."

James wrote in his notebook: *Institutional courage is the new frontier.*

Lessons from Other Frontiers

Later, at an observatory in Chile, James compared notes with an astrobiologist. They talked about the early days of the search for life beyond Earth, when even NASA dismissed it as science fiction. “Everything begins as heresy,” the astrobiologist said. “Then technology catches up.”

The comparison struck James. Astrobiology had found its anchor through data, including spectra, exoplanets, and chemical signatures. UAP studies would also need its own data renaissance. Without reproducible evidence and academic structures to support it, the field would drift like the lights it studied.

The Classroom Fire

Months later, James visited a freshman philosophy class at a Midwestern university. The day’s topic was *epistemic humility*. The professor projected the Navy’s 2004 “Gimbal” video on the wall.

“What should a scientist do with this?” she asked.

A student raised his hand. “Analyze it like any other anomaly.”

Another responded, “Or ignore it until there’s more data.”

The professor smiled. “Both answers are correct. Science grows by doubt, not dismissal.”

James realized that somewhere between the skeptics and the believers, education had begun to reclaim its purpose, aiming to ignite curiosity rather than extinguish it.

A Future Still Forming

By the time James compiled his notes for publication, the academic landscape had changed noticeably. New symposia were being scheduled at respected universities. Small grants had appeared under neutral labels like "aerospace safety anomalies." A few journals had opened special issues on "unknown aerial events."

The progress was incremental, but it was progress, nonetheless. A field once treated as career poison was slowly turning into an opportunity for interdisciplinary exploration.

For James, the real breakthrough was philosophical. Academia was beginning to recognize that studying UAP was not about chasing mystery but about testing the limits of human knowledge.

The Light Over the Library

On a spring night, James walked across the quiet quad of a university where he had once been told never to mention UFOs in a lecture. A light moved across the sky, perfectly steady, crossing from horizon to horizon. A student beside him pointed up. "Do you think it's one of them?"

James smiled. "Probably a satellite." He paused, then added, "But that doesn't make it less interesting."

The student nodded, watching the light disappear beyond the library roof.

For James La Pierre, academia's slow embrace of UAP studies mirrored that moment perfectly. Curiosity tempered by caution and skepticism balanced by wonder. The universities had not solved the mystery, but they had finally admitted that the question itself was worth asking.

PART IV

LAW, POWER, AND THE OPEN SKY

"The price of freedom is eternal vigilance."

- Thomas Jefferson -

By the time James La Pierre entered the policy corridors of Washington and Geneva, the UAP story had moved far beyond radar traces and eyewitness reports. What began as unexplained lights in the sky had become a question of law, diplomacy, and responsibility. Governments, private companies, and civilian investigators were now grappling with how humanity should respond to the unknown.

Part IV traces this shift from observation to governance. It examines emerging legal and ethical frameworks as aviation safety rules, transparency laws, and international bodies begin to confront UAP as a legitimate policy concern. Alongside these efforts, civilian research groups, aerospace firms, and open-source networks are building parallel systems of data collection and analysis that rival official programs. Together, these developments show that UAP governance is no longer confined to states alone. It is a shared and contested space where authority, accountability, and curiosity intersect, forcing society to decide not only what is being observed, but who bears responsibility for understanding it.

THE COVENANT OF THE SKY

"Ethics is nothing else than reverence for life."

- Albert Schweitzer -

The Closed Hearing

James La Pierre waited outside a Capitol Hill hearing room, his notebook tucked beneath a stack of government documents labeled Restricted Distribution. Inside, members of Congress questioned defense officials about unidentified aerial phenomena detected near restricted airspace. The terms aviation safety and transparency reform were spoken in the same breath, a rare convergence of law, science, and politics.

He realized this was the new frontier of the UAP debate. It was no longer only a question for pilots or astronomers, but for lawmakers and ethicists. Governments were finally being forced to acknowledge that these anomalies, whatever they were, had legal and moral consequences. As the doors opened and staffers streamed out, James stepped forward, ready to trace how a subject once confined to rumor was becoming a matter of international governance.

The Sky and the Law

In a quiet hangar at Reagan National Airport, James interviewed Captain Elaine Harker, a veteran commercial pilot with thirty years of flight experience. She leaned against a maintenance cart. Her eyes were focused on the distant runways shimmering in the summer heat.

"Pilots see strange things more often than people think," she said. "But there's no real place to put that information. You report a bird strike, a laser, or a drone, fine. But mention something that doesn't fit the list, and suddenly you're the story."

Her tone carried both resignation and irritation. James knew that under the Federal Aviation Administration's guidelines, pilots were required to report any anomaly that could affect flight safety, yet the system had no explicit category for UAP. Reports often ended up buried in vague "unidentified hazard" entries, stripped of context.

Later, at an ICAO workshop in Montreal, aviation regulators from multiple nations echoed Harker's concerns. They spoke about "reporting blind spots" and the fear of ridicule that discouraged accurate documentation. A French official mentioned GEIPAN, France's civilian UAP investigation bureau. "We found that by taking the stigma away," he said, "we improved safety. Pilots stopped whispering and started reporting."

In Santiago, Chile, James met researchers from CEFAA, a civil–military organization that maintained one of the most transparent public UAP archives in the world. They showed him case files where radar operators and flight crews had logged synchronized observations. "It's about discipline," one investigator told him. "You don't need to believe in anything to care about safety."

The message was consistent across continents. UAP incidents, though statistically rare, demanded standardized reporting protocols. Without them, regulators were flying blind. For James, aviation law had become the quiet battleground where credibility and caution intersected.

The Paper Wall of Secrecy

Weeks later, James sat in a cramped office at a Washington law firm that specialized in Freedom of Information Act litigation. Boxes of partially declassified documents surrounded him, filled with thick black redactions. The attorney, a FOIA veteran named Laura Roberts, tapped a stack of appeal letters.

"These are all UAP cases," she said. "Every one of them denied or delayed. The pattern is predictable. invoke national security, redact the details, and stall until the story fades."

She handed James a folder containing the now-famous "Tic Tac" encounter documents. "It took ten years to get this much," she said. The pages revealed less than they concealed, yet the small fragments that survived redaction were enough to confirm that pilots had indeed witnessed something extraordinary.

James compared the American approach to models abroad. In France, GEIPAN's online archive offered full reports and analysis. Brazil's Information Access Law had opened its Air Force archives to the public. Even the United Kingdom, once secretive, had declassified thousands of Ministry of Defence UFO files. Yet in the United States, transparency came only through pressure, lawsuits, or leaks.

At a conference in Geneva, transparency advocates argued that secrecy had become the greatest obstacle to understanding. A policy analyst from Chile on the panel phrased it sharply. "Secrecy breeds myth. Transparency breeds accountability." James noted how often that tension appeared, with governments citing national security while citizens demanded truth. The lack of a unified global framework left each country improvising its own balance between disclosure and defense.

The Ethics of the Unknown

The ethical dimension of the UAP question drew James to Oxford University, where a small group of philosophers and astrobiologists met to discuss "non-human intelligence." The room smelled of coffee and old books. On a whiteboard, someone had written, *"If not alone, then what?"*

Dr. Sophia Vakoch, a bioethicist, spoke first. "If we ever confirm that UAP represent non-human intelligence," she said, "the first ethical question isn't how to respond. It is who gets to respond. Who speaks for humanity?"

Another scholar compared the issue to bioethics and artificial intelligence governance. "We already deal with entities that challenge our definition of personhood," he said. "The same principles apply. Transparency, accountability, and respect for autonomy, even across unknown forms of life."

James recorded every word. The discussion soon shifted to planetary protection, the principle that humanity has a responsibility to avoid contaminating or exploiting alien ecosystems. One scientist suggested that the ethical question might someday be reversed if contact occurred on Earth instead. "What if we are the ones being studied?" he asked. "Would we want our culture distorted or our autonomy taken from us?"

After the session, James walked through the college courtyard, replaying the discussion in his mind. Ethics, he realized, was not a luxury reserved for philosophers. It was the framework that would determine whether humanity approached discovery with humility or hubris. In an age when governments debated weaponization of new technologies, ethical preparation was not an abstraction but a necessity.

The Policymakers' Dilemma

In Vienna, at the United Nations Office for Outer Space Affairs, James attended a roundtable with delegates from fifteen countries. They spoke in cautious diplomatic tones about "anomalous aerial incidents" and "shared skies." The conversation quickly revealed how fragmented the global response remained.

A U.S. delegate emphasized national security. "We must protect classified technologies," she said. A French representative countered, "Transparency builds trust." The Brazilian delegate proposed a hybrid model with civil-military cooperation under UN oversight. The chair of the meeting, an Austrian diplomat, summed up the challenge. "We are attempting to govern something we do not yet understand."

James watched as representatives from ICAO and COPUOS debated whether UAP should fall under aviation safety or space governance. Both sides had valid points. The phenomena crossed boundaries of airspace, jurisdiction, and imagination. Traditional legal categories no longer fit.

Later that evening, in the lobby bar of a nearby hotel, a European policy analyst told James quietly, "This issue is going to define international law in the twenty-first century. It forces us to ask whether the sky belongs to nations or to knowledge itself."

That thought stayed with him. Governance was no longer just about sovereignty. It was about how humanity handled uncertainty on a planetary scale.

The Policy Architects

Back in Washington, James attended a closed-door workshop convened by aerospace lawyers, scientists, and defense officials. The task was to design a draft framework for UAP governance. One group worked on aviation safety, another on data transparency, a third on ethical and international policy coordination.

Dr. Raymond Pelton, a space policy expert, addressed the room. "We can learn from space law," he said. "The Outer Space Treaty taught us that no nation owns the heavens. UAP policy should extend that logic. Shared sky and shared responsibility."

An engineer proposed anonymized global reporting systems for pilots, like the ICAO's accident databases but inclusive of anomalous encounters. A sociologist argued for citizen participation through open data platforms. "Public trust," she said, "is the currency of legitimacy."

By the end of the session, a consensus had formed. UAP governance required cooperation, not competition. It needed to integrate aviation law, scientific transparency, and ethical foresight under a unified framework. It was, as one participant said, "a chance to prove that humanity can respond to the unknown without fear or denial."

The Ethics Briefing

James was invited to observe an ethics simulation at a defense think tank. The scenario imagined a near-future event where a verified craft of unknown origin entered Earth's atmosphere. The participants were divided into teams, including military, scientific, diplomatic, and media response units.

The exercise quickly exposed the fractures. The military team wanted containment. The scientists argued for open observation. The diplomats insisted on a joint UN statement. The media group leaked partial information before confirmation. Within two hours, the simulation descended into chaos.

At the debrief, the moderator turned to James. "You've spent years studying this. What did we learn?"

He paused before answering. "That our first contact will not be with another species," he said. "It will be with our own systems, our laws, our institutions, our fears. How we manage that will define whether we face the unknown with panic or with principle."

The Road Ahead

Months later, James compiled his findings into a comprehensive report titled *Sky, Law, and Conscience*. It outlined reforms for aviation safety, transparency, ethics, and international coordination. Among his recommendations were new UAP categories within ICAO's reporting systems, revised FOIA guidelines to reduce over-classification, and the establishment of an international ethics council under UNOOSA.

He sent copies to policymakers and academic colleagues. Some replied with enthusiasm, others with polite silence. Change, he knew, would be slow. But the walls were beginning to move.

That evening, as he walked past the Lincoln Memorial reflecting pool, the city lights shimmered across the water. The reflection reminded him of the dual nature of the UAP debate, a dialogue between the real and the imagined, the visible and the hidden.

For James La Pierre, the story of UAP governance was not about conspiracy or revelation. It was about responsibility. The way humanity responded to its own uncertainty would determine whether the age of disclosure became an age of wisdom or merely another chapter of fear.

The Covenant of the Sky

On his final field visit, James stood beside an air traffic controller at a busy international airport. The radar screen glowed with tracks of aircraft moving in perfect coordination. Then, briefly, a new signal appeared, fast and erratic, before vanishing. The controller marked it and shrugged. "Probably nothing," he said.

James wrote a single line in his notebook: *Probably nothing is where everything begins.*

In that uncertainty, he saw the essence of the challenge. Law, ethics, and policy were not shields against mystery but tools for meeting it responsibly. The question was no longer whether UAP were real, but whether humanity was ready to treat the unknown as part of its shared moral and legal universe.

THE INNOVATORS OF THE UNKNOWN

"Innovation is born at the margins."

- Clayton Christensen -

The Basement Conference

The hotel ballroom smelled faintly of coffee and projection-lamp heat. Folding chairs filled every corner, and on stage a hand-painted banner read *"Citizen Science and the New Sky Frontier."* James La Pierre took a seat in the back row, notebook balanced on his knee. Around him sat pilots, software engineers, retired physicists, and a scattering of UFO enthusiasts wearing conference badges.

A presenter from the Scientific Coalition for UAP Studies was describing a radar analysis of the 2004 *Tic Tac* encounter. The graphs on the screen were meticulous, but what fascinated James more was the mood in the room. A quiet determination that serious research could happen outside government walls.

"Innovation begins when institutions hesitate," the speaker said. "If the agencies cannot share their data, then we must build our own instruments."

That sentence would become James's guiding note. Over the following months, he traveled across laboratories, garages, and boardrooms tracing how the private sector and civilian investigators had quietly built a parallel research infrastructure for studying the skies.

The Archivists of the Unknown

James's first stop was a converted warehouse in Cincinnati, the unassuming headquarters of the Mutual UFO Network. Metal shelves held decades of manila folders filled with sighting reports, sketches, and interview transcripts dating back to the 1970s. A volunteer investigator named Carla, wearing a headset and a MUFON logo shirt, greeted him with the enthusiasm of someone defending a lifetime's work.

"We're not the tabloids people think we are," she said, opening the online dashboard that now replaced the filing cabinets. "Every report gets coded, geo-tagged, and reviewed by at least two investigators. We train people like crime-scene analysts."

On the screen, dots scattered across a world map pulsed with recent submissions. Each dot was a story. A flash over Kansas. A metallic disk above Madrid. A triangle near Seoul.

James asked how many of the reports held up under scrutiny. "Maybe ten percent," Carla said. "But even the noise is data. It tells you where people are watching, what they expect to see, how culture changes their interpretation."

Later, in the quiet of the archive room, James flipped through an early case file typed on thin paper. The sincerity of the witnesses struck him. MUFON, he realized, was less about chasing saucers and more about preserving the raw texture of human observation, a living record of wonder, uncertainty, and misperception.

The Engineers

From Cincinnati, James flew south to Florida to visit UAPx, a group of Navy veterans-turned-engineers. Their modest lab looked like a startup with cables snaking across tables, monitors showing test footage from infrared cameras, and a whiteboard filled with formulas.

Co-founder Matt Powell greeted him with a handshake and a grin. “We’re field scientists with better Wi-Fi,” he joked. Then he led James to a van retrofitted with telescopes, magnetometers, and spectrum analyzers.

“Our goal is simple,” Powell explained. “Real-time, multi-sensor data. If something’s out there, we should be able to measure it across the spectrum.”

James accompanied the team on a night deployment near Catalina Island, a stretch of ocean long whispered about by pilots and fishermen. The van hummed with the sound of cooling fans as laptops logged feeds from cameras and radar receivers. For hours the monitors displayed only stars and passing aircraft. Then, briefly, an object flared on one infrared sensor before vanishing.

Powell replayed the recording again and again. “Probably a thermal inversion,” he admitted, scribbling notes. “But every time we run a test, we refine the system.”

Watching them work, James felt he was witnessing the rebirth of ufology as engineering. Gone were the flashlight-and-tape-recorder days. These investigators spoke the language of data pipelines and error margins.

The Scholars Without Tenure

At an academic conference in Colorado, James met the founders of the Scientific Coalition for UAP Studies. Their panel drew physicists, psychologists, and former defense analysts. Dr. Kevin Knuth, a physicist from Albany, projected mathematical models of acceleration profiles measured in military videos.

"We treat this as an unknown engineering problem," Knuth told the audience. "Extraordinary data require ordinary discipline."

Afterward, over coffee, he confided to James that SCU's greatest challenge was perception. "We can write perfect papers, but peer-review boards hesitate to publish anything labeled 'UAP.' The stigma lags behind the evidence."

SCU's approach impressed James. They had small teams, cross-disciplinary review, and public access to methods. It was academia in exile and building credibility line by line.

The Aerospace Corridor

Las Vegas shimmered in the desert heat as James drove toward a warehouse complex bearing the logo of Bigelow Aerospace. The receptionist led him through echoing corridors lined with mock-ups of inflatable space modules.

A project manager agreed to speak on background. "Our founder believed the UAP issue required aerospace precision," he said. "So BAASS applied structural analysis, materials testing, even medical monitoring to anomalous reports. Some of it overlapped with defense contracts, some didn't."

James sensed the boundary between curiosity and classified research was thin here. The manager smiled when asked about *Skinwalker Ranch*. "Let's just say it taught us that extraordinary environments demand extraordinary instrumentation."

Outside, as jets roared overhead, James reflected that private aerospace firms possessed resources that governments envied, including money, advanced equipment, and freedom from bureaucracy. Yet they also guarded their findings closely, caught between scientific ambition and corporate secrecy.

The Algorithm Hunters

In a sleek Manhattan office, James met the leadership team of Enigma Labs, a tech startup whose walls were covered with real-time maps of global UAP sightings. Rows of engineers monitored clusters of colored dots moving across continents.

“Our users send videos and coordinates through the app,” explained data scientist Mei Leung. “Machine learning filters ninety-nine percent as explainable. The rest gets flagged for review.”

James watched the algorithm score incoming reports by credibility, cross-checking them with satellite trajectories and air-traffic data. It was ufology translated into the grammar of big data.

When he asked whether the company would share its raw datasets, Mei hesitated. “We publish summaries,” she said carefully. “But privacy, intellectual property, and security all matter. Open data sounds good until a competitor mines it.”

Her answer revealed the paradox of modern civilian research. The same profit motive that fueled innovation also constrained transparency. Still, James couldn’t help admiring the scale of their operation. For the first time, sightings from rural Argentina to urban Seoul were entering a single analytical stream.

The Open-Source Dreamers

To balance the corporate perspective, James sought the grassroots networks that prized openness above all. He joined a video call with volunteers from SkyHub, a now-defunct project that had once installed AI-driven sensor stations on rooftops.

On the screen appeared small labs and garages filled with 3D-printed camera mounts and circuit boards. “We wanted anyone to contribute,” said project coordinator Ian Williams. “Our code is still online. Others can pick up where we left off.”

Though SkyHub had folded for lack of funding, its legacy lived on in the UFO Data Project, which James later visited in Oregon. There, scientists and hobbyists were building low-cost optical and radio-frequency detectors that uploaded their readings to open databases. “Every contributor owns their data,” explained engineer Laura López. “But they also agree to share it freely. That’s how trust grows.”

At night, James helped the volunteers align a sensor beneath the clear sky. The work felt communal, almost sacred, like a modern version of amateur astronomy where curiosity itself was the reward.

The Balancing Act

Across his travels, James saw recurring patterns. Innovation alongside fragmentation and courage that gets shadowed by stigma. MUFON’s vast archives suffered uneven quality control. SCU’s rigorous papers reached only small audiences. UAPx needed funding for new expeditions. Startups guarded proprietary code. Grassroots networks struggled to survive.

Yet he also saw opportunity. Partnerships between universities and civilian groups were beginning to form. Aerospace companies were lending sensors for joint field tests. Even government agencies, once dismissive, were quietly monitoring these civilian datasets for cross-correlation.

At a symposium in Boston, James presented his interim findings. “The private sector and citizen science,” he said, “are not substitutes for government research. They are catalysts. They make secrecy impossible to sustain.”

The room erupted in discussion about standards, funding, and international cooperation. For the first time, it felt as though the separate worlds of corporate, academic, and amateur researchers might converge around shared protocols.

The Global Thread

Invited to a conference in Paris organized by the French space agency’s GEIPAN unit, James met delegates from Brazil, Japan, and Chile. They debated how civilian data could complement national archives. A Brazilian representative described his country’s transparency law that released decades of Air Force files. “Public access,” he said, “creates public responsibility.”

James noticed how these global models inspired younger researchers. A team from Argentina proposed a network linking citizen sensors across continents. A Canadian developer demonstrated an AI tool for linguistic analysis of sighting narratives.

By the end of the conference, James understood that the future of UAP study was becoming decentralized, collaborative, and international. What began as isolated enthusiasts had evolved into a distributed intelligence system for watching the skies.

Reflections at Dusk

Back home, James spread his notes across a long table. He drew three circles labeled *Civilian, Corporate,* and *Open Source*, then connected them with arrows of cooperation and tension. At their intersection he wrote: *Credibility through collaboration.*

The phone buzzed with a new message from a contact at UAPx. "Another deployment next month. Want to ride along?" He smiled and typed back, "Always."

Outside, twilight settled over the city. Airliners climbed through the orange haze, their lights tracing deliberate paths toward the stratosphere. Somewhere beyond them, other lights waited, unexplained and unclaimed, still drawing humanity toward curiosity.

For James La Pierre, the story of private and civilian ufology was not only about technology or discovery. It was about the human refusal to leave the unknown unexplored, the determination to measure mystery rather than mythologize it.

In the flicker of data streams and the hum of homemade sensors, he saw a quiet revolution where science was reborn at the margins moving steadily toward the center.

PART V

TOWARD A UNIFIED SKY

"The important thing is not to stop questioning."

- Albert Einstein -

By the time James La Pierre reached the final stage of his investigation, ufology had transformed from a scattered history of secrets and sightings into a single, intricate mosaic. Governments guarding information, scientists demanding proof, citizens refusing silence, and policymakers struggling to keep pace all revealed different faces of the same mystery. What once seemed fragmented now appeared connected, shaped by decades of tension between curiosity and control. Patterns emerged where contradiction once ruled. The mystery gained coherence without losing its depth.

Part V gathers these threads into a unified vision of what the study of Unidentified Aerial Phenomena can become. Looking back on division and forward to integration, this section shows how history, science, culture, and governance converge into one conversation about knowledge. It argues that ufology matures only when these domains speak to one another rather than past one another. The question is no longer only what moves across the sky, but what that pursuit reveals about humanity's enduring need to understand its place in the universe.

THE ARCHITECTS OF CONVERGENCE

"The future depends on what you do today."

- Mahatma Gandhi -

The Long View

James La Pierre stood at the edge of a windswept hill overlooking a decommissioned radar station in New Mexico. The concrete dome, once alive with Cold War urgency, sat silent beneath the setting sun. For decades, sites like this had collected signals, traces, and mysteries that slipped through the cracks of explanation.

He thought of how ufology had mirrored the structure of this landscape. Scattered, isolated, and half-forgotten. Governments had guarded secrets behind closed doors, academics had turned away to protect reputations, and private citizens had gathered fragments of the unknown with little support. Yet something had begun to change. The boundaries that once divided these worlds were eroding, replaced by cautious cooperation and a shared recognition that the skies could no longer be dismissed as empty.

As evening shadows lengthened across the desert floor, James opened his notebook and wrote one sentence that would frame his next report. *The future of ufology will depend on what humanity decides to do together.*

Echoes of Fragmentation

James's journey began in the archives of the National Air and Space Museum, where he pored over yellowed memos from Project Blue Book. The files revealed a familiar cycle. Secrecy followed by speculation and investigation followed by dismissal. Government reports, television specials, and academic critiques all told the same story of fragmentation.

In one folder, an Air Force officer had scribbled, *No further study warranted.* Another, dated a decade later, warned of public panic if certain data were released. The oscillation between suppression and sensationalism had shaped the entire field.

Later, at a small university in Massachusetts, James met Dr. Karen Clancy, a sociologist studying scientific stigma. "It isn't just about the data," she explained. "It's about who is allowed to ask the questions." Her students were analyzing how institutions define legitimacy and how that definition shifts when public curiosity becomes too loud to ignore.

Her words stayed with him. He realized that ufology's future would not come from a single breakthrough or discovery, but from dismantling the barriers that had kept curiosity in silos for over seventy years.

The Five Pillars

James's research began to form around what he later called the Five Pillars of Integration. They became the foundation for a framework he would present to policymakers, researchers, and the public alike.

The first was historical literacy. Without understanding past missteps, from the Cold War secrecy of Project Blue Book to the limited transparency of AATIP, humanity risked repeating cycles of ignorance.

The second was scientific rigor. He met engineers experimenting with multimodal sensors that could cross-verify radar, infrared, and optical data. One physicist compared it to seismology. "You do not call an earthquake an earthquake until at least three stations record the same tremor."

The third pillar was sociocultural awareness. At a conference in London, James listened to anthropologists explain how cultural narratives, including angels, sky gods, and extraterrestrials, were all variations of a single human pattern of projecting meaning onto the unknown.

The fourth pillar centered on governance and ethics. Lawmakers in Washington debated reforms to Freedom of Information processes and pilot reporting standards. At the same time, ethicists were drafting guidelines for hypothetical contact with non-human intelligence, treating it as an extension of planetary protection and AI governance.

The fifth and final pillar was interdisciplinary collaboration. Everywhere he went, James encountered researchers who had stopped asking *Who owns the data?* and begun asking *How can we share it responsibly?*

Each pillar represented a shift from isolation toward integration, creating a framework for a discipline that could one day be recognized as ufological science.

The Architects of Convergence

In California, James visited a joint workshop between the Scientific Coalition for UAP Studies and the aerospace startup UAPx. Engineers calibrated sensors while psychologists tested protocols for witness interviews. On a nearby monitor, a machine-learning model parsed footage from night-vision cameras.

"Each of us used to work alone," said physicist Raul Vega. "Now we design our experiments together. If the Air Force won't share data, we build our own sky network."

The same spirit of cooperation appeared in Washington, where congressional hearings brought defense officials, scientists, and civilian advocates together for the first time. The language was bureaucratic, with terms like "airspace integrity," "anomaly reporting," and "public accountability." Behind the language though, James recognized something unprecedented. It was institutional humility.

Across the Atlantic, France's GEIPAN program and Brazil's open-access archives were proving that transparency and security need not be enemies. The international community was beginning to experiment with models of openness that valued both credibility and caution.

The Human Lens

Despite the growing sophistication of sensors, James knew that the human element remained central. He spent days interviewing witnesses whose lives had been changed by unexplained experiences. Some were pilots who still whispered about encounters near restricted airspace. Others were civilians whose stories had been dismissed for decades.

A retired Chilean radar operator told him, "The instruments tell part of the truth. The people tell the rest."

Psychologists explained how perception, bias, and expectation shaped these accounts. Cultural theorists traced connections between folklore and modern sightings. The phenomenon, James realized, existed both outwardly and inwardly, serving as a mirror that reflected humanity's hopes and fears as much as any external mystery.

In his notes he wrote: *To study the unknown sky is also to study ourselves.*

The Governance Web

At the United Nations Office for Outer Space Affairs in Vienna, James attended a policy session discussing how UAP reporting might be integrated into existing aviation and space frameworks. Delegates from fifteen nations debated clauses that blended law, science, and ethics.

The chairperson from Austria summarized the dilemma. "Our airspace is shared. Our knowledge must be as well."

ICAO representatives discussed pilot safety. COPUOS specialists raised the issue of technological secrecy. James watched alliances form and dissolve in real time. Every delegation wanted transparency in principle but control in practice.

That tension, he realized, would define the next era of ufology. Governance would have to walk a tightrope between openness and protection, between curiosity and caution.

The Builders of the Future

Back in the United States, James spent weeks documenting how private and civilian innovators were reshaping the field. At Enigma Labs, he saw developers refining algorithms that could filter thousands of crowd-sourced reports each day. "We are building the world's largest structured dataset of aerial anomalies," said data scientist Mei Leung. "Our challenge is accuracy, not volume."

In Oregon, volunteers from the UFO Data Project assembled low-cost optical sensors in garages, uploading their readings to open archives. The simplicity impressed him. "This is what democratized science looks like," one participant said.

At a symposium, a MUFON archivist, an SCU physicist, and an AI engineer shared a single stage for the first time. Their differences in language and method remained, but they spoke with a shared conviction that collaboration mattered more than consensus.

A Framework for Tomorrow

James's research eventually culminated in a visual roadmap he presented at a global policy forum titled *From Fragmentation to Integration*. The framework displayed concentric circles representing data integrity, human perception, governance, ethics, and public engagement. At the center was not a UFO but a compass, its needle pointing toward cooperation.

He proposed international protocols modeled after aviation safety networks, standardized metadata for sensor data, and open channels between citizen scientists and institutional researchers. Each recommendation aimed to make the study of UAP less a collection of mysteries and more a practice of responsible inquiry.

During the Q and A, a delegate asked whether he believed the phenomenon itself would ever be fully explained. James smiled. "Perhaps not," he said. "But explanation is not the only goal. The goal is to learn how we look for truth together."

The Broader Horizon

Outside the conference hall, the night sky stretched over Geneva like a black ocean dotted with distant points of light. James reflected on the decades of fear, ridicule, and fascination that had shaped humanity's response to the unknown. The shift he had witnessed, moving from secrecy to dialogue and from isolation to cooperation, felt like the beginning of a more mature curiosity.

He recalled Gandhi's words that opened his manuscript. *The future depends on what you do today.* The phrase no longer sounded abstract. Each conversation, each sensor deployed, each declassified document was part of an unfolding pattern of accountability.

Ufology, he realized, had become a mirror for the entire scientific enterprise. It tested the integrity of governments, the creativity of engineers, the empathy of philosophers, and the imagination of citizens. It was not just about the skies. It was about the human capacity to confront uncertainty with both reason and wonder.

The Light That Persists

Months later, back at the same New Mexico ridge where his journey had begun, James looked up once more. The old radar dome stood quiet, its purpose long faded. Above it, a satellite traced a steady path across the heavens.

He closed his notebook for the final time and whispered a thought meant only for himself. "The mystery never ends, but neither does the effort to understand it."

In that moment, he understood the future of ufology not as a hunt for proof but as an evolving dialogue connecting technology with testimony, science with story, and the known with the unknowable.

And in that delicate balance between skepticism and wonder, the study of the unknown finally began to look like a science of connection itself.

AN UNFINISHED SKY

"We are just an advanced breed of monkeys on a minor planet of a very average star."

- Stephen Hawking -

James La Pierre returned to the desert where his journey had begun, the air thin and electric with the smell of rain. The sun was sinking behind the distant mesas, casting long amber streaks across the sand. The same radar dome he had stood beside months before rose from the earth like an artifact of another age, its curved surface now rusted and quiet. It had once gathered echoes from the sky. Now it gathered silence.

He reflected on how far the field had come and how far it still needed to go. Ufology had always acted as a mirror, reflecting the era that studied it. During the Cold War, it reflected fear and secrecy. In the digital age, it reflected noise and curiosity. Yet in the spaces between those extremes, in the quiet halls of universities, in engineers' laboratories, and in the notebooks of investigators like him, it had begun to reflect something new, maturity.

The patterns he had followed were no longer about proof or belief, but about process. The question was not *"Are they real?"* but *"How do we learn to ask better questions?"* Science was evolving toward transparency, and policy was evolving toward responsibility. Ordinary citizens were becoming part of a global network of watchers, data collectors, and thinkers. The mystery had not disappeared,

but it had changed shape. It now lived as a collective effort, not a solitary chase.
James sat on the cracked foundation of an old instrument platform and watched the first stars appear. Above him, satellites traced slow, deliberate arcs across the sky. He knew that somewhere else, another team of scientists, volunteers, and perhaps even skeptics was observing the same sky from another part of the world, recording data in hopes of making sense of what still eluded them all.

He wrote in his final field notes: "*The unknown is not a void. It is a meeting place. It exists wherever courage and curiosity come together. The task is not to prove or disprove, but to listen to the data, to the witnesses, and to each other, and to continue refining the tools that make that listening possible.*"

As the night deepened, he realized that the story of UAP was also the story of human persistence. Every signal, every misreading, every revelation had carried the same message. That the act of seeking itself is what binds humanity to the cosmos.

When he finally closed his notebook, he felt no sense of ending. The investigation would continue, carried forward by new hands and new approaches. The sky above him, vast and indifferent, remained what it had always been. An unfinished question waiting to be asked again.

And beneath it, humanity kept watching.

LIST OF AUTHORITIES

AAAS (American Association for the Advancement of Science): Occasional scientific symposia addressing anomalous aerial phenomena.

AARO (All-domain Anomaly Resolution Office): Current U.S. Department of Defense office overseeing UAP analysis (est. 2022).

AATIP (Advanced Aerospace Threat Identification Program): Pentagon UAP research program (2007–2012).

AAUP (American Association of University Professors): Advocacy on academic freedom and controversial research topics.

AI Research Consortia (e.g., OpenAI, DeepMind): Developers of machine learning systems applied to anomaly detection and data fusion.

APA (American Psychological Association): Frameworks for analyzing cognition, perception, and belief formation.

APRO (Aerial Phenomena Research Organization): Early civilian UFO research society (1952–1988).

Arizona State University (ASU): Academic seminars linking UAP studies to astrobiology and SETI.

Bigelow Aerospace / BAASS (Bigelow Aerospace Advanced Space Studies): Aerospace contractor conducting UAP-related research under U.S. defense contracts.

Brazilian Air Force (FAB): Military investigations and public releases of UAP archives.

CIA (Central Intelligence Agency): Historical role in Cold War intelligence and disinformation related to UFOs.

Clancy, Karen: Sociologist examining stigma and knowledge production in UAP discourse.

Congress of the United States: Oversight hearings, legislative mandates, and whistleblower testimonies concerning UAP.

COPUOS (United Nations Committee on the Peaceful Uses of Outer Space): International coordination on space governance, planetary protection, and UAP policy.

DARPA (Defense Advanced Research Projects Agency): Development of aerospace technologies and advanced sensor systems.

Enigma Labs: Private-sector data aggregation platform using AI for UAP reporting and analysis.

ESO (European Southern Observatory): Exoplanet detection and spectroscopic studies relevant to technosignature research.

Eurocontrol: European aviation safety agency; harmonization of incident reporting systems.

FAA (Federal Aviation Administration): U.S. civil aviation authority managing air safety and pilot reporting protocols.

FBI (Federal Bureau of Investigation): Early Cold War-era interest in UFO reports and security implications.

FOIA (Freedom of Information Act, U.S.): Mechanism for UAP-related transparency and declassification.

Gallup: Public opinion polling on extraterrestrial life and UAP belief trends.

GEIPAN (Groupe d'Études et d'Informations sur les Phénomènes Aérospatiaux Non identifiés): France's national UAP research division under CNES.

ICAO (International Civil Aviation Organization): Global air safety guidelines and standardized reporting frameworks.

IAU (International Astronomical Union): Coordination of global standards for astronomical observation and data classification.

LANL (Los Alamos National Laboratory): Research in advanced materials and isotopic analysis.

MIT Lincoln Laboratory: Development of radar, satellite, and data-integration technologies.

MOD (UK Ministry of Defence): Official investigations and public release of historical UFO files.

MUFON (Mutual UFO Network): Civilian organization collecting and investigating UAP sightings worldwide.

NASA (National Aeronautics and Space Administration): Space sciences and aerospace collaboration on UAP-adjacent phenomena.

NASA Exoplanet Exploration Program: Studies of habitable exoplanets and atmospheric biosignatures.

NATO (North Atlantic Treaty Organization): Military alliance assessing defense implications of anomalous aerial objects.

NICAP (National Investigations Committee on Aerial Phenomena): Mid-20th-century civilian UAP research group (1956–1980).

NOAA (National Oceanic and Atmospheric Administration): Provider of atmospheric and satellite observation data.

NORAD (North American Aerospace Defense Command): Bi-national U.S.–Canadian system monitoring airspace and aerial incursions.

NRO (National Reconnaissance Office): Operator of classified orbital reconnaissance assets relevant to UAP detection.

Pew Research Center: Sociological research and surveys on belief in UFOs and extraterrestrial life.

Project Blue Book (U.S. Air Force): Official government UAP investigation program (1952–1969).

Sandia National Laboratories: Aerospace materials testing and sensor research.

SCU (Scientific Coalition for UAP Studies): Interdisciplinary scientific organization promoting peer-reviewed UAP research.

SETI Institute (Search for Extraterrestrial Intelligence): Radio and optical searches for extraterrestrial technosignatures.

Skeptical Inquirer / CSI (Committee for Skeptical Inquiry): Promotion of scientific skepticism regarding anomalous claims.

SkyHub: Former open-source UAP monitoring initiative using AI-driven sensor nodes (2020–2022).

Stanford University: Academic research on anomalous materials and isotopic anomalies linked to UAP.

To The Stars Academy of Arts & Science (TTSA): Public–private initiative promoting disclosure, entertainment, and research.

UAPTF (Unidentified Aerial Phenomena Task Force): U.S. Department of Defense successor to AATIP (2017–2021).

UAPx: Independent research organization conducting field-based, multi-sensor UAP observations.

UFO Data Project: Citizen science initiative developing open-access monitoring systems for UAP data.

UN (United Nations): Multinational policy body addressing potential coordination on UAP phenomena.

UNOOSA (United Nations Office for Outer Space Affairs): Facilitator of space law and coordination on international UAP governance.

Kepler and TESS Missions: Space telescopes advancing exoplanet discovery and characterization.

www.ingramcontent.com/pod-product-compliance
Lightning Source LLC
LaVergne TN
LVHW011656100826
845155LV00004B/9

9798218943592